CUTTING FOR CONSTRUCTION

CUTTING FOR CONSTRUCTION

A handbook of methods and applications of hard cutting and breaking on site

David Lazenby and Paul Phillips

The Architectural Press, London
Halsted Press Division
John Wiley & Sons, New York

First published in 1978 in Great Britain by
The Architectural Press Ltd

ISBN 0-85139-130-3 (Architectural Press)
ISBN 0-470-26437-3 (Wiley)

Published in the U.S.A. by
Halsted Press, a Division of
John Wiley & Sons, Inc.,
New York

British Library Cataloguing in Publication Data
Lazenby, David William
Cutting for construction.
1. Building—Tools and implements
2. Cutting
I. Title II. Phillips, Paul
690 TH153 78-040610

Printed in Great Britain by
BAS Printers Limited, Over Wallop, Hampshire

CONTENTS

PREFACE

This book is concerned with cutting tough building materials, for purposes as diverse as total demolition or dismantling and mere surface treatment. The majority of applications will lie between these extremes; the greatest interest will probably be focussed on controlled demolition or part-removal of structures, and their modification by cutting in order to permit re-use or introduction of new structural features.

The modern techniques for carrying out these operations differ from the time-honoured and familiar procedures in two principal respects, firstly, by reducing manual labour, and secondly, by introducing precision in place of crudeness (with the obvious implications for subsequent making-good). Of course, in dealing with so varied a range of processes, many other benefits will be apparent, particular to the technique or the situation. Some techniques have opened up completely new industries, by enabling work to be done which had hitherto been impossible; diamond sawing and drilling, with its application to extensive post-cutting of doorways, windows, service ducts/holes, etc, in concrete structures is typical of this trend.

The cutting and breaking tools and techniques which form the subject of this study are widely available in most industrialized countries. However, they still remain fringe activities of large industrial companies, or the province of the small entrepreneurial operator, and therefore little attempt has been made to publicize advanced cutting methods as a whole. The diamond processes, with publicity and promotional support of the powerful diamond industry, are probably the only exception; their success and market dominance are substantially due to this support. By comparison, the flame processes are much less well known or appreciated, and the other techniques are almost unknown beyond the confined circle of practitioners and a few customers. We will deal in considerable detail with each of the principal techniques of both cutting and breaking; diamond tool processes, a number of heat/flame processes, and water-jet cutting, as well as modern applications of more traditional methods, such as rock drills and explosives, and a variety of breaking techniques.

The designer or specifier should be enabled to identify not only the techniques which are capable of dealing with the work where traditional methods are likely to fail altogether, but also the techniques which bring advantages in cost or time, or minimize disturbance and reduce risks. Thus he must be aware of the immediate technical aspects of the tools and processes, and consider the support services and facilities needed, the hazards, and the environmental effects of each. With all this background information, he can then draw comparisons between whichever processes are potentially useful. We believe that there is a real need for a study which compares objectively all the current techniques and processes, both with each other and with traditional methods, and covers the other aspects which are required for such a comparison to be helpful.

Architects, engineers, builders, developers and plant/maintenance engineers are among those who are likely to need recourse to such a reference at some time, whether to solve a particular problem, or to reassure themselves that they can assess and monitor whatever is being offered by others; it is important for all parties involved in such a project to be fore-warned and fore-armed. After a review of the technical aspects of the processes, we turn to a discussion of their potential, and have included a number of case studies with the intention of stimulating better and more imaginative use of what is available. Thus, in seeking both to inform and to stimulate, we hope that this book may help to promote a more active and competent industry in the UK and elsewhere.

We have drawn on knowledge and experience of the UK and other countries, particularly North America; a good deal of the photographic material and many of the case studies come from the USA. We have received help and encouragement from many quarters, which we gratefully acknowledge.

David Lazenby
Paul Phillips
June 1978

ILLUSTRATION ACKNOWLEDGEMENTS

Arizona Concrete Cutting & Coring Co 2.14, 3.3
Architects' Journal 3.7, 5.3
Battelle Centre de Recherche de Genève 10.1, 10.2
BOC Ltd 2.8, 2.13, 2.17, 2.18, 2.19, 2.20, 2.22, 2.24, 2.27, 2.28, 2.29, 2.30, 2.31, 3.8, 4.3, 5.1, 6.1, 6.3, 6.5, 6.12, 6.13, 8.3, 8.13, 8.14, 9.1, 9.2, 10.3
Building Research Establishment 10.5, 10.6, 10.7
Cement and Concrete Association 6.14
Clearway Chasing Ltd 2.23
Concrete Cutting Industries Inc 2.21
Concrete Drilling and Sawing Co (Chicago) 2.15, 6.10, 6.4a and b, 6.11, 8.6, 8.8, 8.9
Concrete Sawing & Drilling Association (USA) 2.6, 2.7, 2.10, 2.11, 8.2, 8.7
Consolidated Pneumatic Tool Co Ltd 1.1, 3.1, 3.2, 5.2
Controlled Demolition Inc 8.15
Rubert Dorgan Contractors Ltd 3.5, 3.6
Errut Ltd 2.1, 2.2, 2.3, 2.4, 8.1
Robert G. Evans Co (Target) 2.9
F. A. Hughes Ltd 2.34, 2.35, 2.36, 6.2
Hydrostress A.G. 2.16
Hymac Ltd 3.4, 3.9, 3.10
Sir Robert McAlpine & Sons Ltd 8.10
Messer Griesheim 4.2
Edward Monti 2.32, 2.33
Smithsonian Institution, Washington DC 1.3
Thames Water Authority (with BOC Ltd) 8.4
Tomagest of Liechtenstein 2.25, 2.26, 4.4, 9.3a and b
Trollope & Colls Ltd (with the *Daily Telegraph* and Kaybore Ltd) 8.12
Western Coring & Equipment Co 2.12
George Wimpey & Co Ltd 8.11

1 INTRODUCTION

1.1 HISTORICAL BACKGROUND AND DEVELOPMENT

Improvements over the last two or three thousand years have seen the development of percussive tools and techniques to cut and form building materials, first stone and then later man-made materials including concrete. These historic methods evolved from, and around, the hammer and chisel for cutting the hard materials, and from early knowledge of the manner in which friction could be applied to cutting in order to make sawing a practical method for the softer stones. Prior to the introduction of machinery, improvement came from the development of more effective methods of applying these cutting mechanisms, and the introduction of harder and more durable tools. However, they were obviously slow, required considerable manpower and were increasingly unable to cope with the modern materials of the last 150 years.

In comparison with the technological changes in some other industries, there is nothing startling about the nature of much present-day equipment, and its introduction on a significant scale has taken a long time. This is largely due to the low level of demand that has existed until recent years. Without a doubt it was the advent of reinforced concrete structures and the subsequent changes in the methods of building that have accelerated the development of more sophisticated in situ cutting methods, and the advanced procedures possible today.

1.1 Heavy pneumatic concrete breakers being used for a typical, tough secondary breaking job

The earliest patent relating to the construction of loadbearing slabs with iron rods or wire ropes in tension was that taken out in England by W. B. Wilkinson in 1854. Although this probably pre-dates the work being done in France, and America, the real exploitation of the material was left to the French engineer François Hennebique. The imported French system was then brought to Britain about 50 years later. The oldest surviving reinforced concrete building in Britain is now Weavers Mill at Swansea, constructed about 1897, one of Wilkinson's buildings, nearly 50 years older, having been demolished in the early 1970s.

The rapid expansion in the use of reinforced concrete for buildings that started after the First World War did not, however, immediately generate the need to cut or break in any different manner. Construction was highly labour intensive, and there was then no incentive to seek mechanical solutions. Furthermore the very fact that concrete was a "formable" material probably diverted attention away from the eventual need to cut and shape it after hardening, which only became apparent when those first buildings required adaptation or demolition. As one example of this slow acceptance of new techniques, it took over 30 years from the first introduction of diamond tipped cutters and thermic lances on to the construction site for them to be widely adopted commercially.

The present range of equipment used in the construction industry

Table 1.1 The development of cutting technology

Historic antecedent	Present method	Future development
Battering ram	demolition ball	
Hammer and chisel	concrete breaker rock drill tungsten-carbide "flail"	
Hammer and wedge	hydraulic burster gas-expansion burster	
Gunpowder	explosives	
Quarrying	diamond drill diamond saw	
Metal working	thermic lance powder cutting	plasma torch
Mining and stone working	rocket jet burner	
Industrial/manufacturing technology		micro-wave cutting laser eddy-current heating

for breaking and cutting has evolved in general along one of two paths, sometimes from origins outside the building trades. The first is by the improvement of traditional manual tools by mechanisation, and the second is by the introduction of completely new techniques. The former route led to the pneumatic breaker, the steel demolition ball, and the rock drill, whilst the latter applies to the thermic lance (or burning bar), the diamond tipped cutter, the water jet, and the other more recent and advanced concepts, eg other methods of thermal energy. Table 1.1 illustrates the development of the various methods of cutting and breaking.

1.1.1 PRIMARY AND SECONDARY OPERATIONS

It is important to understand the distinction between primary and secondary operations. "Primary cutting" is the direct activity of cutting the building member, either to remove or alter it. The redundant material or the debris from this operation must then be removed. Small debris can be dealt with in a variety of ways, but large items (eg panels of wall or floor which have been removed, sections of a building which have been separated from the remainder) may well need further breaking up to facilitate their disposal, and this subsequent work is called "secondary breaking".

1.1.2 TRADITIONAL METHODS

Methods involving tools to cause attrition of the building material must have the longest history, having evolved from flint tools, through the hammer and chisel, to the sophisticated equipment of today; certainly in more recent years, diamonds, discussed further below, have played a most important role in the development of these techniques. The second group, covering splitting or bursting methods, was a technique taken to a fine art by the early stone masons, as a method for quarrying accurately sized stones. The labour of hammering in wedges into a prepared hole lined with a split steel tube, called feathers, has been superseded by hydraulic power, and the significance of this technique now lies mainly in its effectiveness for demolishing mass concrete or brickwork.

1.1.3 INNOVATORY METHODS

Diamond cutting

Diamond is the hardest mineral known to man, having an "indentation hardness' value some 15 times that of quartzitic material, and it is this property which has led to its use as a cutting material (see the graph shown in **1.2**). Fortunately it is also fairly abundant in the grades used industrially, and its cutting properties have been appreciated for a long time.

In Tolansky's book on the history and use of diamonds, he quotes from Diderot's *Encyclopaedia* published in 1751, describing the method used by quarry workers for preparing shot-holes in hard stone. The holes were bored with a heavy iron bar with a cutting face of diamond points, which was allowed to impact on to the surface under its own weight. Water was used to wash away the rock slurry. More recognizable as antecedents of our present-day drills would be the equipment devised by Leschot in 1862 for rock drilling, and later put to the test in the construction of the Mont Blanc tunnel.

Diamond crown bits for rock drilling continued to be improved and were easily capable of penetrating the hardest rock. They were occasionally used against mass concrete on major civil engineering sites by engineers with experience of their capabilities. However there were few noticeable landmarks in the development of diamond tools for the construction site until after the end of the Second World War. Probably the first breakthrough came as a result of the work done at Oak Ridge National Laboratory, Tennessee, in 1948. Faced with the problem of

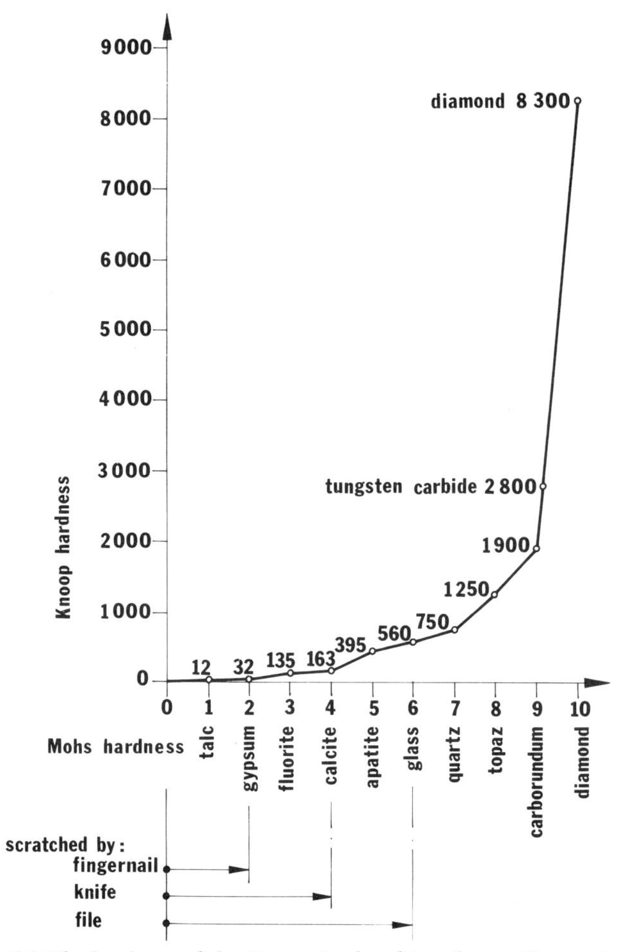

1.2 The hardness of the diamond related to other cutting materials

1.3 Rock drills from the 1876 Centenary Exhibition on a reconstruction of the stand at the Smithsonian Institution, Washington

taking test cores from the concrete shield of a nuclear reactor, they successfully applied diamond core bits to reinforced concrete. Considerable difficulty was experienced with the steel reinforcing bars, and two of the three holes had to be repositioned. Similar cores were taken again in 1958, by which time the steel drilling problem had been overcome.
The early cutters were incapable of cutting steel reinforcement until about 1955. In the United Kingdom, the Atomic Energy Authority contributed a great deal towards the improvement of coring bits for drilling heavily reinforced concrete through the work done at Windscale and Calder Hall Power Stations from 1953 onwards. Test coring at Calder Hall could not be done using water as the flushing and cooling medium, and a technique using compressed air was successfully employed.
Diamond processes, and the equipment to apply them, were becoming able to tackle a wider range of work. With increasing concern for environmental factors, such as noise and fumes, it is likely that the relative merits of these processes will become even more significant in the future.

Rotary sawing

Rotary sawing had been familiar in the timber and metal working trades for many years before it became widely employed as a method for cutting stone. Circular saw blades with diamond cutting surfaces are understood to have originated in France in the stone industry in 1898, after many years of development by a M. Fromholt following a period during which mechanised reciprocating stone sawing was widely used. Early attempts at using diamonds to improve the cutting capabilities of these machines probably started with the introduction of a diamond-bearing paste between the steel saw blade and workface. Its application to other building materials, particularly concrete, would have obvious merit, but of course necessitated a different technique, namely in situ cutting with portable machines. All these early saw blades were mounted on large fixed machines to which the work had to be brought for cutting to size prior to erection. Thus, in the late 1940s work was being done in several places to prove that in situ concrete sawing was feasible. By 1947 Jim Doyle, working in Canada, had produced the first wall saw and 1949 brought the first recorded use of diamond sawn contraction joints in pavement concrete at Topeka, USA. The pioneer of the road planer and bump-cutter, Cecil Hatcher of California, carried out his first successful contract at Davis Airforce Base, Tucson, in 1952.
So far as machinery is concerned, it is interesting to note the current reversion to electric and hydraulic power transmission systems after a long dependence on compressed air for both sawing and drilling. The benefits of having far more useable power at the cutting tool have overcome the initial advantages of air, which were its universal availability and simplicity.

Flame processes

The technique of passing a controlled flow of oxygen down a length of mild steel pipe which had been heated to ignition temperature (about 900°C) and using the heat of the resulting reaction to melt otherwise difficult masses was first described by Thomas Fletcher to the Society of the Chemical Industry in Liverpool in 1888. Burning the mild steel pipe on its own would produce insufficient heat to melt non-ferrous or non-metallic masses. This shortcoming was later overcome by packing the steel tube with wire, the oxygen continuing to flow down the tube in the spaces between the wire rods. The heat generated by this increased mass of iron being consumed by the oxygen was now sufficient to melt concrete, forming a fluid iron silicate slag which was constantly blown away by the oxygen. This became the technique of thermic boring which was improved in France following the Second World War, and was employed to speed the demolition of massive German fortifications.

Other methods

Other methods of thermal energy cutting were tried in the late 1930s, notably in the work of an Englishman living in America, Robert Aitchison, who was employed by the Union Carbide Corporation. He proved the effectiveness of adding aluminium powder to a post-mixed oxygen/acetylene flame to produce higher temperature flame than had previously been possible. From his work evolved the powder lance, the powder torch, and the rocket-jet piercing-burner. This type of equipment has been progressively developed so as to improve its compactness and manoeuvrability and present-day equipment is dealt with in detail in a later chapter (pp. 19–30).

1.2 PRESENT EXPLOITATION OF CUTTING METHODS

The historical background to the most common techniques leads on to the principal streams of specialist cutting operations which apply at present. The most common processes are those of mechanical attrition (usually implying diamond tools), impact/explosives, and heat processes. Lesser activities are water-jet cutting, and specialist secondary breaking.
Of course all these co-exist with the so-called "traditional" methods, such as the pick, the pneumatic breaker, the rock drill and wedge burster, and the crashing simplicity of the steel ball or bulldozer. Jobs frequently require that a number of different techniques are all used in a carefully determined sequence, which can call for considerable understanding of each of the processes and good planning so as to obtain the maximum benefit.
The market for specialized cutting techniques, both primary and secondary, is now fairly well served throughout the industrialised countries. There are many companies specializing in diamond drilling and sawing, thermic lancing, or the application of explosives, and there is a wide range of equipment available for purchase or hire by contractors who wish to use their own work-force. In addition, there are particular specializations, such as surface treatment of concrete, which serve to concentrate expertise in some small companies.
Unfortunately, the optimum use is rarely made of all the techniques that are available because they are so often brought in only to rectify faults, or to attempt to restore a project to programme after traditional methods have failed. It is to be hoped that, in the future, engineers and designers will make an informed decision as to the methods to be used, based on better knowledge, and will pre-plan the appropriate cutting into the programme.
To some extent the present ad hoc employment of specialist cutting firms results from the comparatively high costs which appear to be involved, and from the fragmented, dare we say discordant, attempts of the various suppliers of services to communicate with their potential market.
Poor communication is also to blame for the fact that projects involving the removal of redundant or unwanted parts of a structure, where the latest cutting and breaking techniques would be beneficial, are sometimes abandoned because the technical difficulties and the assumed costs are seen as too much of a barrier. The successful exploitation of these techniques will increasingly call for the accumulation of engineering and design experience at the professional level.
However, a good deal of reliable information is available to cover most aspects either from manufacturers, and in Britain from organizations such as the Building Research Establishment or De Beers, publishers of *Industrial Diamond Review* (for diamond

applications); in the USA, information is obtainable from the Industrial Diamond Association or the Concrete Sawing and Drilling Association. The information available varies widely and this, together with the rather haphazard way in which the industry has grown in Britain, makes it essential for the specifier to be well informed, so that he can judge the relative merits of the various methods and act accordingly.

In the USA the level of knowledge is not necessarily greater, but the specialist contractors are generally much better organized and equipped, and therefore economic operation can be expected as a matter of course. As a result of progressively improving the economic performance of diamond cutting against that of burning, the flame processes are used relatively infrequently, and diamond processes are the standard approach in most cases, in the USA.

A further factor retarding the adoption of most modern techniques is the difficulty of matching the complexity and engineering delicacy of the equipment with a correspondingly capable workforce. Recognition must be given to the fact that operations such as diamond sawing or powder flame cutting involve the use of very expensive machines, and demand a high level of operator skill if they are to function economically. The skills required are predominantly those of an experienced technician, to which there needs to be added an appreciation of building technology. This is certainly not the context in which to employ casual labourers, without any training, as too often has been the case.

1.3 THE ROLE OF THE STRUCTURAL ENGINEER AND THE SPECIALIST CONTRACTOR

When the designer has established his objective, he must beware of the pitfalls which await those who do not take some obvious precautions. Specialist advice is likely to be required in two respects, first to assess the structural implications of the alterations proposed, and secondly to verify the technical aspects of the cutting processes themselves.

Experience has shown that structural engineering expertise is often essential, and even when it is not, it can be beneficial. The structural engineer will probably have been a member of the design team concerned with new construction, or refurbishment schemes; he may be called in to advise on particular aspects of more minor alterations.

The steps to be followed in correctly establishing the specialist work are:

(i) Assessment of the structural implications of the work in respect of the structural integrity of the building

The more obvious work is in checking the effects of removing all or part of structural members, or ensuring that holes may be formed without prejudicing the structural integrity of the members. Particularly when dealing with existing buildings, he will need to look out for special structural forms which are more susceptible to weakening by cutting operations, eg prestressed concrete (particularly unbonded post-tensioned members), compositely designed steel beams/concrete slabs (which may not tolerate removal of areas of slab), reinforced concrete shear walls (which may be resisting horizontal forces as well as carrying vertical loads), and so on.

(ii) Identification of the materials likely to be encountered

A certain amount of structural detective work may also be helpful, in pre-identifying the type of material likely to be encountered, eg cast-iron columns rather than steel, breeze—or rubble—concrete in filler joist floors. Time spent in early investigation will often be more than justified in avoiding time-consuming and expensive delays to the work if unforeseen difficulties arise.

1.4 A component of a precast concrete floor that has split along a reinforcing bar because of the method used to form the hole for the services pipe

Beware of making assumptions as to the true nature and function of members after only a superficial examination. Examples of items which can be misleading are concrete columns which are actually steel stanchions with concrete casing; reinforced concrete slabs which are filler joist constructions incorporating steel joists; and brick walls or piers which conceal steel or concrete columns.

When the material has been identified, this may dictate the choice of cutting method to be used. However, it is more usual to find that the materials are not the sole, or even major, factor affecting the decision on the technique. The objective of the project and the constraints of the site often feature more prominently in the consideration. Chapter 6 gives guidance on the relevant processes for each situation.

(iii) Identification of the constraints applied by the site, the building or the environment

In addition to the design input, when the actual cutting operations are being assessed, it may be helpful to have a specialist contractor's assistance for the more difficult situations. He should be able to advise on the alternative methods available, the type of plant, site services and environmental aspects associated with each and the amount of time likely to be needed for the work. The following chapters cover the various aspects of the work, and chapter 7 in particular deals with specification and quotation in some detail.

1.5 Loadbearing surface of a precast floor component dangerously reduced by poor workmanship. The correct choice of method could have eliminated this risk

(iv) Assessment of the suitability of alternative methods to meet points (i) and (ii), and their relative economy

From the advice given by the structural engineer and the specialist contractor, alternative methods and their costs can be adequately assessed. It should be noted, however, that while some specialist firms are able to offer a wide range of machinery appropriate to a particular technique, as well as alternative procedures, others are not. Therefore a true comparison between the various methods may not be easy to make, and the advice of those specialist contractors with only a limited capacity may well be unreliable.

1.4 THE ROLE OF THE DESIGNER

The designer is interested in these processes in as much as they enable certain objectives to be achieved more quickly and economically than by traditional methods. In addition, and most importantly, in some cases they bring objectives within the bounds of practicability, when they would otherwise be impossible. Chapter 6 reviews the range of operations where these processes will be essential or helpful, and gives guidance on the corresponding cutting and breaking techniques. The designer can make best use of them, and thereby best achieve his objective, by taking account, at all stages of his work, of the possibilities which are available.

At the conceptual stage there is enormous scope for exploiting the possibilities of removal or adaptation of parts of a building, whether it is a new or existing construction. As the detailed design proceeds there will probably be many instances where a knowledge of the specialist skills can help to overcome a detail problem. When construction is in progress, the designer should be able to initiate the correct use of such skills, or at least be able to assess and monitor them as they are employed by the contractor. All too often the methods are introduced too late to help solve unforeseen problems or to effect remedial work.

The applications are relevant to new construction, covering the formation of openings for services, cutting movement joints for thermal and other reasons, safety grooving in floors and roads, holes for fixtures and fittings, cosmetic treatment and many other aspects. They are possibly even more relevant to building refurbishment, being an essential part of many building alterations, in addition to those aspects which are common in new building. Refurbishment usually implies more effective use of the space within the existing building envelope, and possibly adding to that floor space, together with associated upgrading of the environmental services. Thus the work may include forming lift shafts, stairwells, and service ducts, creating new door and window openings, and removing inconvenient walls or columns which have been made redundant by local strengthening of the structure. In addition to being aware of the range of possible techniques, and their relevance to any particular situation, the designers/specifiers need to have an idea of the real cost of achieving their objectives and to be able to relate the cost of any one method to the alternatives available. Tables 6.4 and 6.5 give very approximate guidance on the relative costs of the most common techniques, relative to the actual cutting operations themselves; the work of other researchers is also quoted in chapter 6.

However, the designer or specifier needs to appraise each situation carefully, because although it is possible to make a comparison between the direct costs of different techniques of cutting, some methods produce better results than others. For example, diamond sawing will produce a dimensionally accurate, neat, regular opening, whereas pneumatic breaking may show a lower cost but will produce very approximate, rough and irregular results. In addition, site restrictions, speed or environmental considerations may impose further limitations on the choice. Thus the characteristics of each process, as discussed in chapters 2 and 3, are relevant to the designer's use and acceptance of them.

1.5 COMMERCIAL ASPECTS

The development of more efficient and convenient equipment and the increasing awareness of the usefulness and adaptability of the techniques have led to an active industry in the USA, Britain, and Europe. However, the larger markets of the USA have enabled their diamond-based procedures to become better organized and equipped than elsewhere.

Some operations are capable of being carried out by a competent general contractor, for which purpose he will hold or hire suitable tools. Others require particular skills and equipment so that they are inevitably the province of specialist contractors. Within the field of such specialists there is a wide range of sizes of firm, technical ability, and scope of operations; in this matter, as in so many others, the old maxim of "horses for courses" holds good. A client, whether a general contractor or a specifier, will be best served by a specialist who is experienced in the particular operation, bearing in mind that some of the larger firms cover all normal methods, and have sufficient capacity to deal with the job as promptly as is necessary. In chapter 7, table 7.1 gives some guidance on the types of specialist contractors likely to be encountered, with an indication of the skills and capacity they may offer.

A more knowledgeable market, coupled with the efforts by specialist firms to build up teams of experienced operators, with a good range of equipment, should lead to a more prompt and economical response to enquiries for such work. The contractual relationships with these specialist contractors is a matter of considerable importance, as also is the problem of understanding the operating costs of the various processes. It is unfortunately impossible to quote detailed cost comparisons between alternative processes for a common objective; in addition to the difficulties involved in finding a basis for such a comparison, and the innumerable variations of circumstances which could affect such an exercise, in the final analysis the best comparison is between properly specified quotations from contractors who are prepared to undertake the work. However, chapter 6 gives broad guidance on common situations.

2 PRIMARY CUTTING

The actual operation of making the cut in the building material, whether to form holes or large openings, remove or adapt a member, or to modify the surface, can be carried out by a wide range of processes and equipment. From the historical review in chapter 1, it will be seen that the processes commercially available at present fall into the following broad categories:

2.1 "Soft" cutting tools, covering abrasive discs, floor grinders, tungsten carbide tipped tools and "flail" floor grinders;

2.2 "Hard" cutting tools, effectively meaning diamond tipped tools, and covering diamond core drills and diamond saws;

2.3 Heat/flame processes, covering oxygen cutting, powder cutting, thermic lance, Thermit powder and flame spalling (oxy-acetylene or oxy-kerosene rocket jet); and

2.4 Water-jet cutting.

Each category will be reviewed with a note of typical methods of application, and their respective advantages and disadvantages. A further summary of the alternative means of achieving certain common objectives is given in chapter 6.

2.1 "SOFT" CUTTING TOOLS

2.1.1 GENERAL NOTE: RANGE AND USES

Operations to saw, drill or chase building materials may be carried out using a variety of tools. The choice of tool will be influenced by the hardness of the material to be cut, and the amount of cutting involved. Concrete, stone, brick, ceramic tiles and even reinforced concrete can be cut with abrasive discs, and drilled or chased with tungsten carbide tipped tools. These cutting tools are soft in comparison with diamond and they are therefore defined as "soft" cutting tools.

Although the efficiency of a cutting edge composed of diamonds (defined as a hard tool) is greatly superior to any other, soft tools, such as abrasive discs or tungsten carbide tipped tools, will function quite satisfactorily on the relatively soft materials involved. Their comparatively short cutting life may well be offset by lower blade costs in appropriate circumstances. An additional advantage is that these are generally tools which can be put into the hands of less skilled operators because the effects of misuse will rarely prove as costly as those of damaging diamond blades or drills. The range of soft cutting tools and their characteristics are illustrated in Table 2.1.

These tools are considered under the following headings:

2.1.2 Abrasive discs;

2.1.3 Floor grinders;

2.1.4 Tungsten carbide tipped tools;

2.1.5 "Flail" equipment.

2.1.2 ABRASIVE DISCS

There are many commercial varieties of abrasive disc on the market, most of which are specially formulated to suit the material which it is intended to cut. The basic disc consists of a fabric base which is impregnated with a hard resin or vitreous mixture, containing grits such as aluminium oxide, silicon carbide (corundum) or diamond. The disc is sometimes reinforced with glass fibre or metal to provide additional stiffness and wear resistance.

Discs are usually run dry, but some types may require water for lubrication, and as a safety precaution to keep down potentially hazardous dust (see chapter 5). They are available in a range of sizes up to a maximum of about 500mm in diameter, but are rapidly consumed during use, and the size correspondingly diminishes. The expected disc life when dry-cutting a 12 × 12mm groove would be a length of about 9m. The rate of output for the same job should be around 330mm/min (1ft/min) with a peripheral speed at the edge of the disc of 10m/sec (equivalent to about 600 rpm); allowance must be made for the time taken to replace worn discs. The machines on which abrasive discs are used are generally lighter and simpler than their counterparts for

Table 2.1 Soft cutting

Process	Applications	Degree of skill	Characteristics
Abrasive discs	quick cutting jobs on pipes, etc. and for chasing or grooving	low	easy to handle but noisy and produce dust. Blades relatively cheap but rapidly consumed
Floor grinders	removing surface irregularities	low	treated surface is ready for applied finishes and may be alternative to screeds
Tungsten carbide and other wear-resisting alloy-tipped tools	small drills, chisels, breaker points and scabbling tools for general use	low/medium	widely available but performance and suitability depend on circumstances
"Flail" surfacing equipment (tungsten carbide tipped)	surface treatment of slabs by reduction and grooving	medium	a special application of tungsten carbide tips. Fast. Treated surface is skid resistant. Suitable for local and medium areas of traffic way, but not extensive highway grooving

Note Generally no preparation is required

2.1 An abrasive disc joint-cutter being run with water cooling and lubrication to the blade

2.2 As an alternative, suitable abrasive discs may be run dry

2.4 Tungsten carbide flail grooving

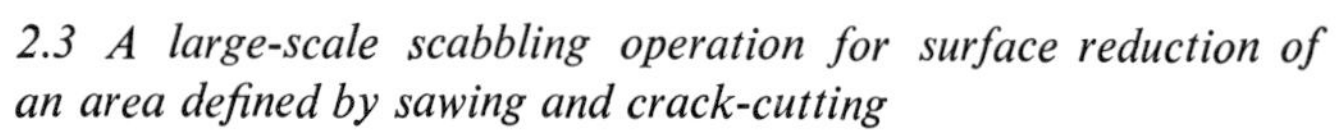

2.3 A large-scale scabbling operation for surface reduction of an area defined by sawing and crack-cutting

diamond blades, although several types of machine are suitable for both.

Applications

Small hand-held saws, to take up to a 350mm (14in) diameter blade, are commonly used for quick cutting jobs on concrete pipes and brickwork, and to form small chases, particularly in walls. While they are light and easy to use, they tend to be noisy, and to produce a good deal of dust. As compared to machines using diamond blades, the lighter weight and lower cutting pressure of the self-propelled floor cutters using these blades have made them popular for cutting contraction joints in "green" concrete, particularly in industrial floors; this technique does not offer the accuracy of diamond-sawn joints in concrete which has reached an age of 24 hours or so.

2.1.3 FLOOR GRINDERS

Abrasive cutters of a different form are used in one design of concrete floor grinder. This device uses abrasive blocks mounted on a pair of heavy contra-rotating metal discs, on a machine with a similar action to an industrial floor scrubber.

Applications

The method can sometimes be used to eliminate the need for finishing screeds by providing a surface smooth enough for the final floor covering. The slab is ground at an early age, usually within 3–4 days of placing, and the resulting surface should be levelled to a degree satisfactory to receive vinyl tiles or 10mm wood blocks applied directly to it.

The purpose and action of this machine are not directly comparable to those of the diamond-bladed bump-cutter, which is generally intended for large-scale levelling on traffic surfaces such as roads and runways, frequently with the intention of leaving a grooved or roughened surface.

2.1.4 TUNGSTEN CARBIDE TIPPED TOOLS

Tungsten carbide is a very hard alloy which can be welded on to the wearing or cutting surfaces of a softer metal tool to extend its life, and to provide a cutting edge suitable for hard abrasive materials. Tungsten carbide tipped tools, especially the small diameter masonry twist drills, are familiar, everyday tools on most building sites. The points used in concrete breakers and tools for surface modification by bush-hammering or scabbling are also of this category. A rotary and percussion action is most frequently employed because of its efficiency. The percussion action alone causes the desired attrition, while the rotary action constantly changes the point of impact, and in the case of drilling, assists in preventing the bit from jamming.

Applications

Masonry twist drills, used in rotary or rotary percussion tools, and cruciform tip rock drill bits will drill most building materials, but they are not suitable for penetrating heavy steel reinforcement or mixes containing large, hard aggregates.

Rock drilling bits may be fitted to specially threaded extension rods to drive holes up to 6m (20ft) in suitable materials. Rock drilled holes can be used economically in conjunction with thermic lances to cut the cost of punching lines of weakening holes, lancing only being necessary when the steel is encountered. They are also widely used for producing holes for starter bars and dowels, for bore holes for explosives, carbon dioxide (Cardox) charges or hydraulic bursters.

2.1.5 "FLAIL" EQUIPMENT

In addition to its use in drill bits and chisels, and for scabbling tools, tungsten carbide has recently been applied in a novel manner for grooving and levelling concrete floors. The cutting principle developed for a range of machines marketed by Errut Products Ltd uses tungsten carbide tipped flails. In this machine, the flails (or tines) are suspended from five planetary shafts which revolve about the main shaft at 1550 rpm. The tines are held extended by centrifugal force and flail the surface with great rapidity to produce continuous grooves. The tines are claimed to last between 3½–5 hours, depending on the qualities of the concrete. The largest unit presently available is a ride-on machine powered by a 22·4kW (30 bhp) petrol engine, and with the capacity for treating 25–30m^2/hr.

Because the work is carried out dry, this machine has an advantage over methods requiring water, when working on airfields or other extensive sites with limited supply points.

2.2 "HARD" CUTTING TOOLS

2.2.1 GENERAL NOTE: RANGE AND USES

The impressive accomplishments resulting from the application of diamond tools, a small selection of which are featured in the case studies, are evidence of the greatly increased scope that they offer to the designer. Only diamond edged tools are capable of continuously machining reinforced concrete, in the sense that machining is an engineering operation, implying accuracy and control over the quality of finish in addition to simply cutting or penetrating.

However, a great deal of the expansion in the use of diamond tools has arisen because of its substitution for other methods, in situations where these attributes alone might have been unimportant, but where the other advantages of minimal vibration, absence of dust and fumes, and economy in meeting the objective have proved decisive factors leading to their use. Table 2.2 illustrates the characteristics of hard cutting tools and other factors influencing their use.

The performance of the tools, and hence their economy, depends on two principal factors, the output of work and the wear that takes place. In turn these are influenced by the skill of the

Table 2.2 Hard cutting (diamond tools)

Process	Applications	Preparation required	Degree of skill	Characteristics
Core-drills	forming holes and extracting core samples	machine set-up	medium	smooth, accurate holes to considerable depth
Floor saws	grooves or openings in slabs	nil	medium	smooth, accurate cutting with high output. Machines generally large
Wall saws	grooves or openings in vertical members	machine and track set-up	medium/high	smooth, accurate cutting
Reciprocal saws	openings through slabs in any plane	machine and track set-up	medium	smooth, accurate cutting to greater depth in one pass. Blade must pass *through* slab

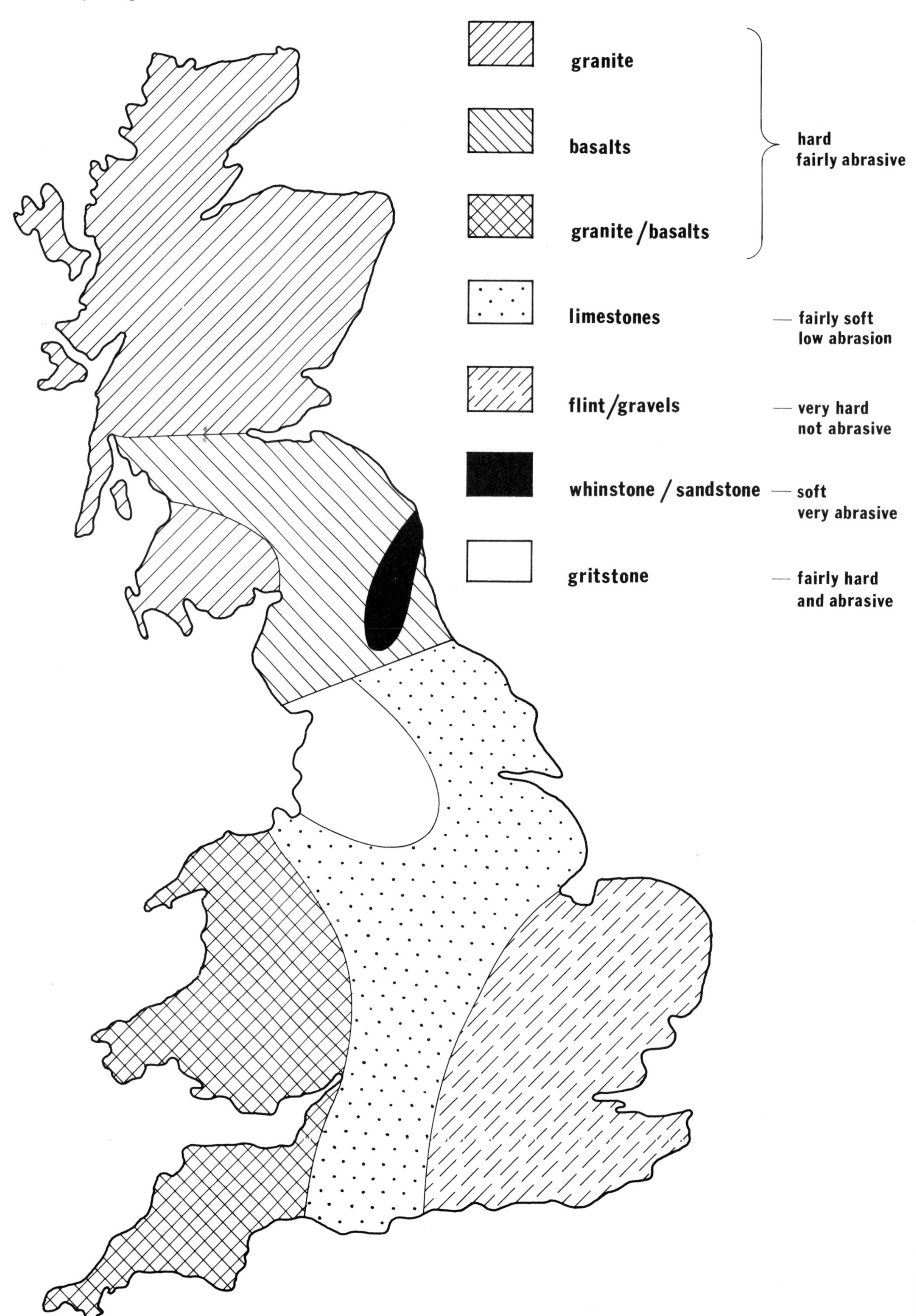

2.5 Varieties of aggregate in the UK

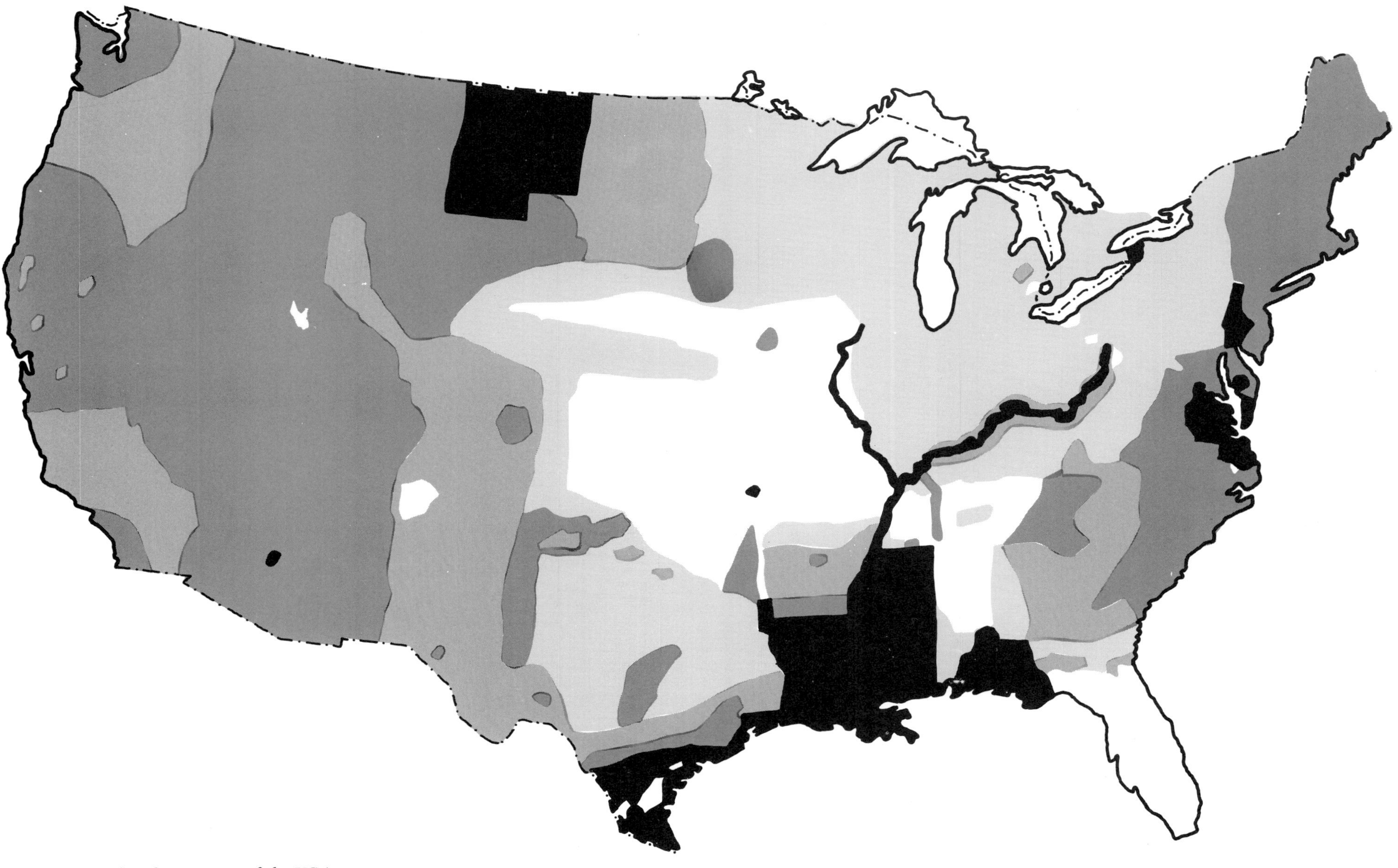

2.6 Aggregate classification map of the USA

operator, the power and condition of the equipment, and the properties of the material being worked.

The successful use of diamond tools is particularly influenced by the competence of the operator, both in terms of output and tool life. While this will obviously affect the economy of the operation, and hence be reflected in the quoted price for the work, nonetheless it is in the specifier's interest to approach only competent and experienced specialist contractors, who are more likely to have available the right men and adequate equipment.

Variations in the material being cut will affect costs in several ways. Chapter 7 discusses how the hardness of the aggregate in concrete can be expected to increase blade wear. The steel content of reinforced concrete will also reduce blade life, by as much as 25 per cent in many cases, and will also slow progress. The type of sand used in making the mortar will also play an important part because the abrasiveness will have the effect of keeping the diamond bearing matrix "open", thus preventing the surface from glazing over. At the same time, however, it will increase total tool wear, and a combination of coarse (large sized) aggregate and abrasive sand can easily halve the normal blade life.

Thus, one of the main problems with diamond cutting is that it is slow and costly when working in concrete with hard aggregates or high steel concentrations.

Concrete, whether it is reinforced or not, is produced in an amazing number of variations. The map on p. 10 gives an approximate guide to the types of rock found in the UK which are commonly used as concrete aggregates. The map on p. 11 shows corresponding information for the USA, based on the experience of members of the Concrete Sawing Committee of the Industrial Diamond Association of America. The sand used in the mix may be the same aggregate crushed, or it might well have been brought in from another part of the country, and could have the effect of completely changing the cutting characteristics of the concrete. Very large flints, of the kind often found in old concrete, have a considerable adverse effect upon diamond drilling, and a slightly lesser effect upon sawing. This is most noticeable when the flint is larger than the diameter of the drill and a stage can be reached when the entire cutting surface is working in a uniform flint groove. Flint is extremely hard, but not abrasive, and there is consequently nothing to keep the diamond matrix open, and thus the segments "glaze over" in a short time. When this happens, cutting virtually ceases, and the bit has to be withdrawn and the segments "dressed down" (a trade term for the operation of exposing fresh diamonds).

Shot-loaded concrete is sometimes encountered in radiation screening in atomic energy power stations and in research establishments. The shot may be of lead or steel, and its purpose is to add to the density of the concrete to improve its screening properties. It is difficult to predict how a particular mix will cut, because it depends upon the other materials with which it is in association. Lead shot will sometimes act as an indestructible lubricant between the diamond and the worked surface. Steel shot behaves differently, and may break out of the matrix of the concrete and roll round under the cutting edges.

Lean-mix or other weak concrete has the effect of breaking up under the pressure of the cutting edge. The aggregate tries to roll around the cutting interface, chipping the diamond segments, and possibly scoring the walls of the bit or saw blade. The erratic load is also detrimental to the machine, and may result in stripped gear trains or broken shear pins.

Very high steel concentrations, as for example in bank vault construction (where steel joists or even old railway lines may be used for reinforcement), can render diamond cutting totally impractical, and flame processes are the only practical solution for cutting objectives, and explosives for demolition. Even when investigation confirms that diamond processes are appropriate, sawing along the line of a reinforcing bar can be a source of trouble. A very slow rate of progress is inevitable because of the constant contact with metal, but a point may be reached when the steel is divided down a major part of its length and one half becomes detached from its bond to the concrete. This usually happens to the thinner of the two halves, and there is then a tendency for it to be drawn into the cut by the movement of the blade. The thin sliver of steel will act as an effective wedge, jamming the blade in its rotation, and probably making it difficult to withdraw from the cut.

Taking all these factors into account, the blades and bits are designed to provide a reasonable compromise to the objectives of high output, and a useful length of life under the expected working conditions. The tool designer will, however, be working to meet an average set of conditions, and the factors mentioned previously may have been excluded from his calculations, which may give rise to a very disappointing performance if they are not noticed before work commences.

Diamond processes are almost always "wet" ones, using cooling water. The amounts of water involved vary between 45–70 l/hr for the average drill, to around 1500 l/hr for a large wall saw. The water problem can be controlled to a large extent by the use of simple water collection methods or by a vacuum water recovery system.

2.2.2 INCORPORATING DIAMONDS INTO CUTTING TOOLS

It was noted earlier, in the historical review, that diamond is by far the hardest and most wear-resistant mineral known to man, and it is these properties which have led to its widespread use as a cutting material. Not only is it available naturally in the grades used for such purposes, but in recent years industrial diamonds have increasingly been synonymous with synthetically produced material; organizations such as De Beers have developed manufacturing techniques to produce industrial diamonds with qualities at least as good as the natural stones, and of a more predictable shape and strength.

In principle, there are two ways of forming a cutting surface to be used for the purpose of working concrete and other very hard building materials. These are by surface-setting the diamonds, or by impregnating solid material such as bronze or steel so that the diamond grits are dispersed as in a matrix.

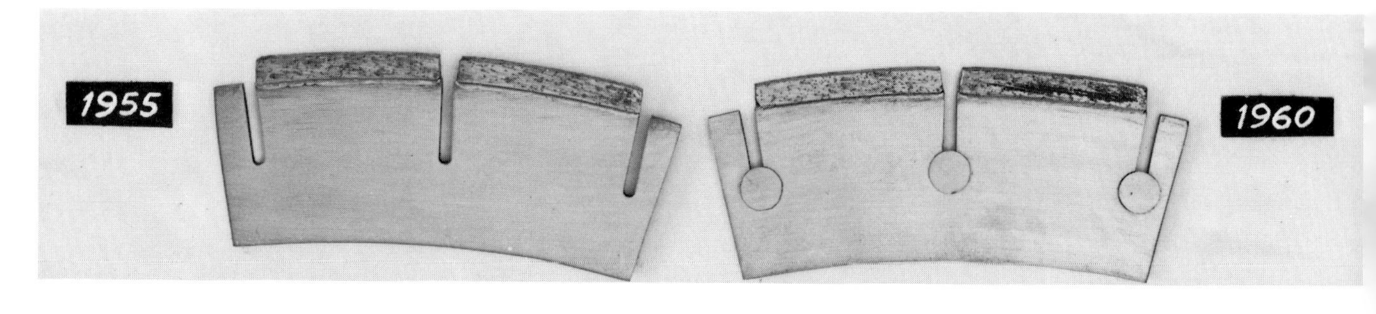

2.7 Cutting edge arrangements of blades produced in 1955 and 1960 showing the use of impregnated matrix segments, brazed in position

2.8 Vertical diamond core-drilling with a compressed air powered machine

Surface-set diamond cutting edges consist of relatively large industrial diamonds set in a bond which is intended to hold the diamond in its exposed position for the maximum cutting life. Used in this way, the function of the bond is to provide a rigid base for the diamond, and it is not intended to be worn away. There is in fact only one superficial layer of diamonds, and the life of the edge is finished once this has worn away.

The diamond impregnated matrix, on the other hand, wears away as the cutting proceeds. It must have a wear rate designed to be compatible with the abrasiveness of the material being worked so that fresh diamonds are constantly being exposed right through the thickness of the cutting edge, or segment of the blade. At the same time the diamonds must not be lost prematurely from the cutting edge before they have exhausted their useful life.

The majority of diamond tools used in the construction industry, both drill bits and saw blades, are of the impregnated type, consisting of carefully graded diamond grits (such as De Beers' EMB or SDA) uniformly distributed throughout the metal base or bond. This material is formed into small blocks or segments which are then attached to an appropriate blank to complete the saw blade or drill.

Behind the apparent simplicity of these tools lies a considerable investment in research and some very clever engineering. This is not always matched by equal skill and care in using the end product, in spite of the fund of experience that is readily available from the leading tool manufacturers. Technical backing for the producers of saw blades and drill bits comes from organizations such as the Diamond Research Laboratory in Johannesburg, who are constantly looking for ways to improve the qualities of the synthetic and natural diamonds used in these tools, and whose work is regularly published.

The manufacturers of such tools are faced with a number of conflicting aims. Cutting segments consisting of impregnated material are manufactured by heating a mixture of metal powder and diamond particles in a graphite mould, to the temperature at which the metal melts and the mixture coalesces (sinters). The diamonds may then be simply held in the resulting mixture by mechanical containment in which case, as the stones become exposed by use, they will quickly reach a point where the grip is insufficient and they will be stripped off the cutting edge. This situation can be improved by creating an affinity between the diamond and its bond by allowing some interchange of chemical components. The diamond producer is therefore being asked to devise a material which will have the size and shape to cut well, but at the same time have the shape to remain mechanically locked into the matrix for as long as possible, and the ability to withstand the high sintering temperatures and the chemical bonding without losing its most important properties of hardness and strength. An edge containing industrial diamonds, bonded in such a matrix, is the only type which will cut a combination of the hardest aggregate and high tensile steel reinforcement. Its efficiency in this application may be judged from the fact that it is not uncommon for the diamond contained in a circular saw blade to reduce to a fine slurry a quantity of material 300,000 times its own mass. This efficiency is doubly impressive when it is considered that, owing to the limitations of the bond between the stone and the metal of the matrix, usually less than half the diamond is actually used in doing the work.

It is obviously essential for the contractor to have the most economical tools at his disposal, whether drill bits, saw blades, grinders or others. The diamond tool industry is constantly developing and improving its products so as to achieve this, and ensure that these more sophisticated techniques will become more economical relative to traditional labour-intensive methods.

Specifiers should remember that diamond tools are engineered to fine limits to suit the material they are intended to cut. It is very much in the designer's interest to furnish the cutting specialist with accurate information about the material to be worked. This data will enable the tool manufacturer to design cutting segments that optimize the combination of hardness of matrix, diamond content and size. A typical variant, if it is known that considerable amounts of steel reinforcement have to be cut, is for the tool manufacturer to add silicon carbide to the bond.

There are two types of diamond edged cutting tool in general use in the building industry, the thin walled drilling bit and saw blades. These will be considered separately.

2.2.3 DIAMOND DRILLS

Thin wall diamond bits consist of good quality, wear resistant, steel tube, with impregnated segments brazed on to the annulus at one end, to form the cutting edge. They are usually driven through a threaded spigot at the opposite end which can be adapted to the power output shaft of the drilling machine. Alternatively, the bits may accept an expanding coupling adaptor to transmit the drive. They are produced in a range of standard diameters, with slight variations from one manufacturer to another. In the UK an attempt has been made at standardization with a proposal for a British Standard to cover the range of diameters as well as the adaptors and extension rods. These generally range from a minimum size of 10mm (⅜in) up to 500mm (20in), and in lengths up to 760mm (30in) for deeper holes. Extension rods can be added to standard bits, or in some instances especially long bits can be fabricated.

The diamond segments fitted to drill bits are of similar characteristics to the impregnated segments used in saw blades, and are usually intended to be run with water cooling. This is accomplished by passing the cooling water down the centre of the driving spigot, into the barrel of the drill. The water will then flush outwards through the slots between the segments, carrying the drilling debris back up the outside of the bit, to emerge at the open end of the hole and eventually drain away.

It is worth noting at this point that water cooling is not absolutely necessary, and diamond cutting has been successfully carried out using either compressed air or nitrogen for flushing the debris. The need for such a technique is only likely to arise when the ingress of cooling water into the workface could be catastrophic, as would be the case, for example, when drilling into a heated furnace lining.

Drilling machines for powering diamond bits successfully, under building site conditions, should have the following features:

(a) a means of rotating the bit at the required speed and torque;
(b) a system for applying a constant pressure to the bit as it penetrates the material;
(c) a stable platform to hold the bit steady while it drills.

Mains supply electricity and compressed air predominate as power sources, but the advent of the high cycle generators and hydraulic power packs, for use with the wall saws, has brought about the conversion of many drilling machines to use these power sources. As with the sawing machines, this has brought about a corresponding increase in power and performance, while at the same time enabling the contractor to have an integrated system of equipment. It has hitherto proved practically impossible to meet the theoretical power requirements of the diamond bit within the constraints of a portable machine, since a 100mm diameter, thin wall bit is capable of absorbing 40 bhp (30kW) at a speed of 200 rpm.

Heavy machines directly coupled to small diesel or petrol engines can be used for drilling downwards into pavement concrete for roadways, for the installation of lights or core sampling. Feed mechanisms are generally manually operated through feed

2.9 *A self-propelled floor sawing machine fitted with its maximum sized blade of 915mm diameter*

2.10 *This road surface has been levelled and grooved in a single operation*

screws directly connected to the sliding carriage on which the motor and drill thrust block are mounted. Automatic feed mechanisms are used on high production rate machines which may be necessary when a large number of holes of great depth have to be drilled, for example when stitch drilling through mass concrete. One particular design of machine has a hydraulically powered automatic feed mechanism operated remotely from a control console, thus enabling several drills to be supervised simultaneously by one operator.

The achievement of a stable platform for the drill can be accomplished by weight alone in some instances, or more involved methods may be necessary, such as vacuum hold-down pads, bracing back to scaffolding or mining bars, or even by bolting directly to the workface. Circular cut-outs, to be formed by stitch drilling a large number of interlocking holes, can be accurately positioned by rotating the drilling machine on a radius arm, centred on a substantial anchorage on the work face. A suitable strong central pivot point for this purpose can be improvised by locating a hydraulic burster into a prepared hole, and then applying a partial hydraulic pressure until a secure grip is obtained.

Applications

The type of bit described above will only form a hole in a single continuous drilling operation when the thickness of the material is less than the effective drilling depth of the bit. If it is greater, then there will have to be a number of intermediate steps in which an annular groove is cut, and then the core which is thus formed, and is of course still attached to the parent body of material, must be broken off and lifted out of position before drilling can recommence. Core removal can be difficult and time consuming, and therefore when the volume of work justifies the additional cost, the drilling contractor will use the maximum length of drill that is practical under the circumstances, to reduce or completely dispense with the necessity for carrying out the decoring operation. (There are a number of simple "tricks of the trade" for removing the core which do not rely on the use of specialized tools. The core is usually broken off by driving wedges into one side of the annular groove, and it is then lifted clear with tongs.)

Very rarely, a designer/specifier will be faced with the need to extract an exceptionally deep core without damaging it, in order to examine the state of the material forming a cross section of a structural member. Under some circumstances it may be possible to employ a geological exploration device known as a wire line coring system. In this technique, the coring bit is driven into the workface until it has drilled the maximum length of core for which it has capacity; this could be as much as three metres. The core is held in an inner sleeve which remains stationary while the barrel of the bit rotates in the cutting operation. At a pre-determined point, a mechanism within the bit causes the inner sleeve to rotate with the outer barrel, and by its grip on the core to break it from the parent mass. The core protected by the inner sleeve can then be withdrawn down the centre of the drill barrel to the surface, for examination.

2.2.4 DIAMOND SAWS

The manufacture of circular diamond blades follows similar practice to that employed for conventional machine saw blades of this form, except that the cutting edge is in this case not "teeth", but segments of diamond impregnated material which are applied separately instead of being formed as an integral part of the whole blade. The blades are made from blanks of high quality, abrasion resistant, steel, on to which the segments are brazed. Following the operation to fit the segments on to the blank discs and a subsequent cleaning stage to remove the residue of flux and brazing, the blades are skilfully stressed (tensioned), so that they run true at speed and under load. This operation has the effect of slightly "dishing", or forming a symmetrical hollow towards the centre of the blade, when it is at rest. (For this reason it is not good practice to start up large blades when they are entrained in a deep cut. Until the blade reaches its designed running speed it would be subjected to a considerable side loading which would reduce its working life.)

The cutting segments are always wider than the thickness of the blade on which they are mounted so that they leave an adequate clearance between the walls of the cut and the blade for the escape of the coolant water and debris, and to minimize frictional losses on the sides of the blade. Obviously the effect of a side loading is to wear away the width of the segments on one side, which will eventually lead to wear on the blade disc itself. The blades are normally intended to be cooled and lubricated by a flow of water which is directed on to the edge of the blade as it enters the cut. In addition, the blades are sometimes deeply slotted to increase the water space, and to aid the flow of slurry.

Rotary sawing machines fall into one of two distinctly different groups, characterized by the plane of the workface upon which they are intended to cut; they are thus commonly described as either wall saws or floor saws. *Reciprocating saws* are also available, with a dual capability.

(a) FLOOR SAWS

These are predominantly self-propelled, and self-contained, wheeled machines, powered by industrial petrol engines. The present range varies in size from about 8 bhp (6·0kW) up to 65 bhp (48·5kW) and in weight from 65kg (150lb) to 460kg (1010lb). The larger machines will take a blade of 900mm (36in) in diameter which is capable of cutting to a depth of 375mm (14¾in), in multiple passes of about 75mm per cut, depending upon the quality of the concrete. At this depth, the cutting speed should average about 8–10 linear metres per hour (26–33ft).

Applications

Floor saws are used for precision cutting of grooves or openings in slabs, both at ground level and suspended. The decision on the size of machine to use for a particular job will be influenced by the complexity of the shape to be cut, the area for manoeuvring, and the floor loading which can be safely imposed. Provided that the machine is not too large to comply with the limitations imposed by these factors, economic performance is optimized by bringing in the largest machine, because cutting speed is largely determined by power.

Floor sawing machines are designed to be guided by the operator, and a sighting aid is usually incorporated to enable the blade to follow chalk lines marked out on the floor. In a situation where a wrongly positioned cut would present a major problem, inverted steel troughs can be bolted to the floor to act as guide rails for the wheels of the machine and so make the control of the operation easier. This use of steel channels will also have the effect of spreading the load of the machine over a larger area of floor. When it is desired to follow random cracks in pavement areas, for instance in opening them out in preparation for repair, a specially designed floor saw as marketed by Cushion Cut Inc of California, which incorporates a castering action to the two front wheels, makes it much more manoeuvrable than the normal machine.

Specialized variants of these machines are the multiple grinders and groovers. These have been developed to treat the surface of trafficways such as highways or airport runways in order to

2.11 In one pass a grooving machine cuts precise channels into a section of a highway surface

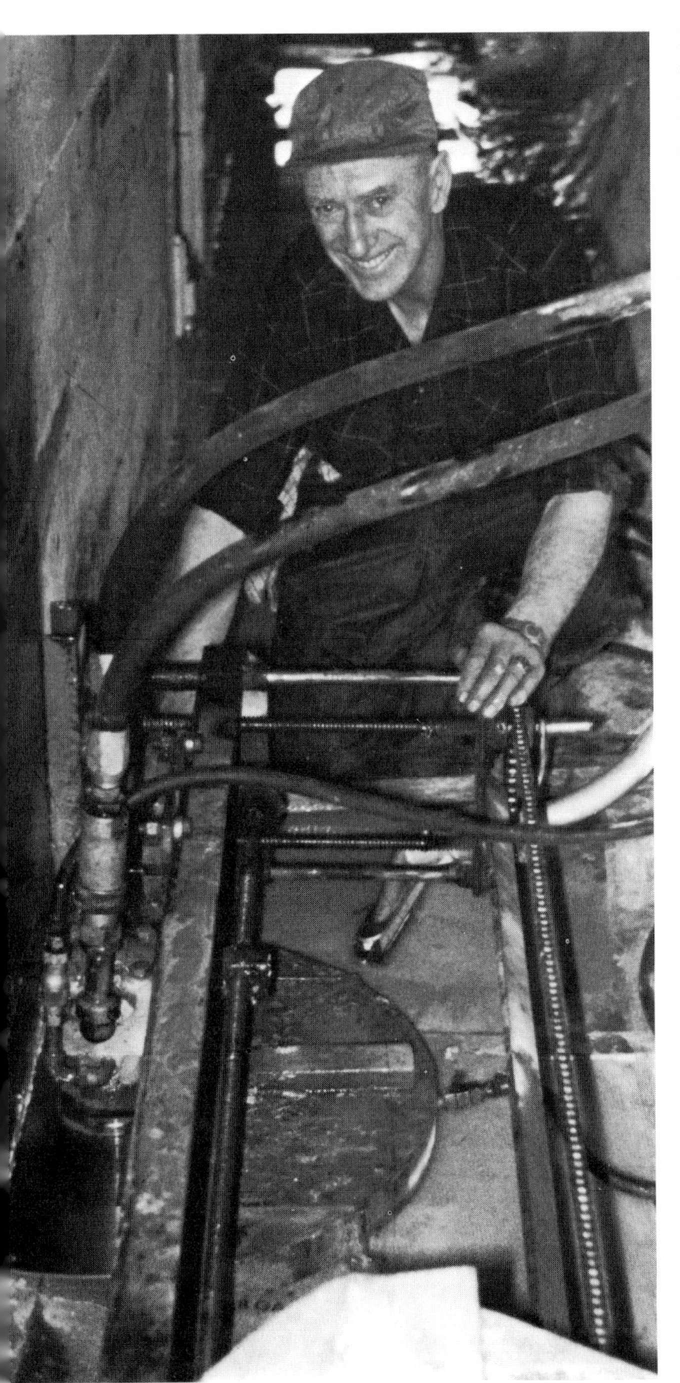

2.12 Sawing concrete in 1947; Jim Doyle's original hydraulic machine

improve their surface alignment, or traction and drainage characteristics. Machines are available in many sizes, up to a width of 4m (12ft) or so, to suit the scale of operations. **2.10** and **2.11** illustrate an out-of-level concrete runway which has been ground to an accurate surface, and a machine fitted with grooving blades at work on a road.

(b) WALL SAWS

The machines which have been designed specifically for cutting vertical surfaces differ radically from those intended for floors. Gravity is generally considered to be an aid to floor sawing because its force acts directly in line with the blade loading, and at the same time stabilizes and provides traction for the carriage. Because wall saws must be lifted into position, and then rigidly supported for their own weight and against the reaction of the cutting forces, the machines are usually made reasonably light by powering them from an external source, via an appropriate transmission system; they traverse the workface on strong tracks, which are frequently bolted to the surface.

Great advances have been made in the sphere of power transmission by improving the power to weight ratio at the point of application. Systems are now available using either compressed air, hydraulics, or electrics (the electrical system may be conventional 3 phase 50 Hz mains supply, or alternatively the new 400 Hz 110 volt, so-called high cycle supply); they transmit the power from remote mobile power sources with diesel or petrol prime movers to the appropriate motors driving the saw blades. The saw motors vary in power from the 7–9 bhp (5–7kW) range of the air motors, to around 35 bhp (26kW) for high cycle or hydraulic.

Compressed air is still the most widely used transmission medium because of the simplicity of the equipment, and the fact that suitable compressors are available on hire, almost anywhere. However there are disadvantages in the use of compressed air for this type of machine, in addition to the low power outputs which they provide. The pneumatic system has a very low overall efficiency, which is often further reduced by icing in cold weather and pipe losses resulting in high running costs, (for instance, a 3mm puncture will waste 0·6m³/min (22cfm)). The exhaust from the pneumatic motors is noisy and not readily silenced.

Both high cycle electric and hydraulic systems are supplanting air because of the way they overcome these problems. The two systems are roughly comparable in performance, but with a number of minor operating differences between them. For example, the cable connections of the electrical system are easier to handle than hydraulic hoses, but the hydraulic system has the advantage of easy speed control, whereas 400 Hz system is constant speed. From the designer's viewpoint, the important observation to make is that the most economical and predictable performance will be achieved by the high power machines, and the quoted cutting prices from most contractors using high cycle or hydraulic equipment will be inclusive of fuel cost, in contrast to compressed air which is an extra cost. When the slow rate of progress and high running costs of the pneumatic system are taken into account, compressor fuel and hire are a significant cost factor. (Each air motor requires a compressor with a capacity of at least 4·5m³/min (170cfm) which will absorb around 65 bhp.)

Wall saws powered by mains electricity supply at 415 volts, 3 phase 50 Hz, are sometimes used in the UK and in Europe, but as the maximum motor size is about 10 bhp (7·4kW) before it becomes too bulky and heavy to manhandle, and there are the risks associated with having high voltage cables trailing around the site, their popularity is limited.

Applications

Tracked wall saws are versatile and accurate machines for precision cutting grooves or openings in walls and vertical members. The majority of designs will operate in any plane except completely inverted, although one model is claimed to be capable of cutting upwards from below a soffit. The tracks are supplied in several lengths (commonly 1·2m and 3·0m), which can be bolted together to give any length required for the job.

The blades used with these machines are in the same range of sizes, and to similar specifications, as those used on the floor saws.

(c) RECIPROCATING SAWS

An alternative to rotary sawing is to employ a reciprocating action, thus duplicating mechanically the movement of the hand

2.13 A compressed air-powered diamond blade saw completing a horizontal cut on the rc wall of a water treatment plant

2.14 A trailer-mounted, petrol-powered hydraulic pump unit suitable for the largest diamond sawing machines

2.15 A typical high cycle, 400 Hz electric generator suitable for powering a diamond saw or drilling rig. The prime-mover is a 60 bhp (45 kW), four cylinder air-cooled petrol engine

2.16 The reciprocating "Hydrostress" diamond blade saw cutting from the required starting hole

saw. This principle is embodied in the "Hydrostress" sabre saw, which has the capacity of cutting reinforced concrete up to 1·2m (4ft) thick, at a single pass, thus greatly exceeding the 600mm maximum depth of the largest rotary blade.

The Hydrostress saw is hydraulically powered from an 18kW (24 bhp) prime mover incorporated in a separate unit with the pump and controls. The machine frame is clamped on to the workface by a highly efficient vacuum system, and a neat gantry is provided as an integral part of the system to take the weight of the machine while it is being moved into position on a vertical surface. The sabre saw blades carry diamond impregnated segments of similar specification to those fitted to circular saw blades and barrel drill bits. As with these configurations, sabre saw blades must be operated with a supply of water to the cutting edge.

Applications

The procedure for cutting an opening is slightly different from that used when circular sawing, because a hole of suitable diameter to pass the saw blade must be drilled before cutting can commence. However, two advantages are gained from the different action; firstly, the sabre saw produces a superior finish to multiple pass circular sawing through thick material, and secondly there is no need to "overcut" to produce sharp corners. The Hydrostress power supply can be teamed with a drilling unit capable of forming the hole needed to start the saw. The manufacturers claim a cutting speed of about 900mm/hr (36in) for a full depth cut through average quality reinforced concrete.

2.3 HEAT/FLAME PROCESSES

2.3.1 GENERAL NOTE: RANGE AND USES

There are several devices which will produce a high temperature flame capable of melting steel (melting point 1600°C), and the silica-containing building materials such as concrete (melting point 1600° to 2500°C). A number of different fuels may be used, which combine oxygen with a fuel gas or other material. The objective is to produce a flame or blast of high temperature gas, capable of easy and close direction, with the correct characteristics to melt and clear away the material from a closely defined target area. Table 2.3 illustrates the application of the heat/flame processes.

The oldest established process is that of oxy-acetylene burning, which is used exclusively for metal working, and from which has developed powder cutting. The best known device to produce a high temperature flame, with the chemical characteristics for melting concrete, is the thermic lance. Two heat processes with more specialized applications are the use of Thermit powder, which is packed around the steel to be cut and then fired to produce a high temperature chemical reaction, and flame spalling, which is a technique for concrete surface treatment using exceptionally large burners consuming oxygen with acetylene or propane.

There are a few cutting problems for which a flame process represents the only answer. These arise when the structure cannot be subjected to the slightest degree of vibration during cutting, or when the steel content is exceptionally high. The flame processes do not impose any mechanical loading on the structure, directly or through vibration, and therefore may be used safely on buildings which may be in danger of uncontrolled collapse as would occur in clearing earthquake or explosion damage. The absence of vibration also means that they are quiet processes, which can be used whenever this is an important consideration. Secondly, the fact that the steel reinforcement melts more easily than the surrounding concrete indicates the manner in which the flame processes become more advantageous as the steel content increases. This may be contrasted with the fact that the presence of large amounts of steel will slow all the alternative cutting methods, possibly to the point where lack of progress and high tool wear make them totally impractical. This is often the case in structures such as bank vaults and wartime shelters in which massive quantities of steel are incorporated in the form of closely laid rolled steel joists, or even old lengths of rail.

Even in less specialized situations, some form of lancing may well be the most effective method for dismantling almost any type of reinforced concrete structure where the steel can be located with a cover-meter and then burned through separately. Used in this way, the minimum amount of costly lancing is used to destroy the major strength of the construction, then enabling a simpler dismantling technique to be used.

Disadvantages mainly centre around the environmental problem of the smoke and fumes, and there is, of course, the fire risk. Also damage may be caused to surrounding areas of the work piece from heat penetration, although this is unlikely to be a problem unless the objective is to create a high strength bond directly on to a lanced surface. If this is the case, then the surface must be mechanically dressed back to sound material to a depth of 25–50mm (1–2in).

Fire prevention starts with an appreciation of the characteristics and risks inherent in the process to be used. The main difference in this respect, between thermic lancing and gas or powder cutting, is the projection of the flame. The latter cutting flame must freely traverse the full depth of the cut, and as a result of this characteristic the flame will usually be visible for a metre or more out from the back face of the work piece, and allowance must be made for this. The thermic lance, on the other hand, usually projects the flame and molten debris back towards the operator while boring, and only emerges at the far side when breakthrough occurs.

All the flame processes cause the formation of a hot slag, the majority of which falls from the cut down to the ground. Some is projected into the working area by the gas pressure, or by spalled fragments and exploding impurities. Precautions therefore have to be taken against projected flames, very hot flying debris (usually no more than sparks, but certainly capable of starting a fire and causing burns to personnel), hot slag underfoot, and not least the radiated heat and hot air currents. Careful examination of the vicinity of the workface must obviously be made, but it is also important to find out if there are any hidden dangers. The ever-present problem of buried services is one such danger, but there may be others such as anti-vibration or temperature-insulation layers in the concrete (for example cork or polystyrene) or interconnecting voids. Other more commonplace hazards are inflammable finishes such as bituminous coatings, any thermoplastics used in false ceilings and light fittings which may melt and drop on to the workface, and scrap timber and rubbish which remains unnoticed until it is well alight.

The quantity of heat retained in the slag is often underestimated with the result that wooden or thermoplastic floors may be found to be damaged although covered with asbestos sheets and a layer of sand. It is therefore sensible to prevent hot slag from accumulating, by periodically cooling with water or removing it from the surface at risk.

The best site conditions for the use of flame processes will obviously be on an open, well-ventilated structure, completely cleared of all combustible material, and evacuated except for the burning team. Anything short of this will demand great care and constant responsible supervision by the contractor, together with the provision of ventilating fans and temporary ducting if the situation is very confined.

The various heat/flame processes will now be reviewed, with commentary on their particular applications and relative merits, under the following headings:

2.3.2 Oxygen cutting;
2.3.3 Powder cutting;
2.3.4 Thermic lancing;
2.3.5 Thermit powder cutting;
2.3.6 Flame spalling.

2.3.2 OXYGEN CUTTING

The operating principle of the oxygen cutting torch is that a pre-heat flame, usually oxygen mixed with acetylene, but possibly with some alternative fuel gas such as propane, or even hydrogen if it is to be used for submerged work, is played on to the surface of the metal to be cut. The intense local heat should be sufficient to raise the temperature at this spot to ignition point, which would be around 900°C for mild steel. A secondary flow of high purity oxygen (better than 98 per cent) is then directed at the heated area, when rapid oxidation should occur. The combustion of the iron to form iron oxide generates sufficient heat to melt the immediately surrounding metal. The high pressure cutting stream blows the molten metal clear of the surface, along with

Table 2.3 Flame processes and water jet cutting

Process	Applications	Degree of skill	Characteristics
Oxygen cutting	ferrous metals	medium	the commonest method for dismantling steel structures
Powder cutting	metals, especially high melting point alloys and silica-bearing materials such as concrete	medium/high	high cost sensitive equipment not suited to rugged open-site conditions
Thermic lancing	cutting and boring most metals and silica-bearing materials such as concrete	low/medium	lances are rapidly consumed and may make up a large bulk and weight to be taken to site for extensive work. Relative advantages depend upon task
Thermit powder	ferrous metal structures	low	an uncommon technique for remotely initiated collapse of steel structures
Flame spalling	surface treatment of concrete for decorative effect or improved traction	low/medium	a quiet vibrationless procedure but with major fire hazard. Best suited to outdoor working
Water jet	surface treatment, crack preparation or cutting concrete	medium	vibrationless procedure with no fire risk. Will not cut steel, so can be used for exposing reinforcement in concrete. Slow and inefficient in use of power

Note Generally no preparation is required

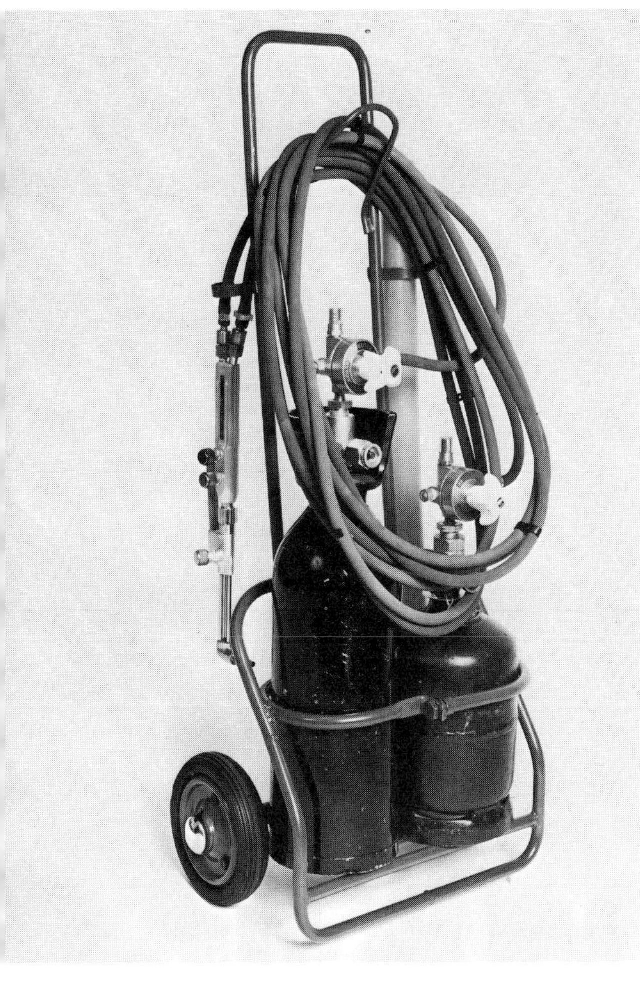

2.17 BOC Portapack uses miniature oxygen and acetylene cylinders. It is ideal equipment to use as a standby in rc concrete breaking or cutting to sever the odd, awkward piece of steel

2.18 A heavy duty oxy-acetylene cutter at work on a 250mm (10in) steel billet

the oxides being formed as the products of combustion.

Two mechanisms for metal removal are therefore at work in successful cutting, the conditions for which are, firstly that the metal must have an ignition temperature higher than its melting point, and secondly that it must burn in oxygen, and combine with any refractory oxide to form a fluid slag. It follows from these conditions that not all ferrous materials can be easily cut in this manner. Among the difficult alloys are those incorporating manganese and aluminium. Cast iron may also be difficult to cut, and will certainly necessitate the use of acetylene as the pre-heat gas, because of its higher flame temperature. The reaction itself will emit a great deal of heat, and additional precautions should be taken to protect both the operator, and surrounding materials.

Although considerable amounts of gas are needed to supply the large torches for extended periods, and this has required large cylinders, there has until recently been no easy way of providing very *small* quantities of gas. The bulk and weight of the commonly available sizes of industrial cylinders (about 1·5m tall, and weighing 100kg) has made contractors reluctant to take them on site unless a considerable amount of work could be foreseen for them. A new development by BOC Ltd called "Portapack" uses specially designed, small, lightweight cylinders for the oxygen and acetylene; these make it practical to have oxygen cutting facilities available on even the smallest job. The complete set of equipment is transportable in the back of a car, and weighs only 25kg (56lb) including the trolley. The cylinders provide sufficient gas to cut 6m (20ft) of 16mm ($\frac{5}{8}$in) thick mild steel; more than enough to "unstick" most reinforced concrete breaking jobs where progress may have been held up by problem steel.

Oxy-acetylene equipment can be operated crudely by almost any site worker following minimal instruction. It is a different matter to use the equipment to maximum effect, for which proper training and considerable experience are required. As with many apparently simple procedures, the difference between the best and the worst standards of operation can be tremendous, and the specifier of a large metal cutting project is advised to check the experience of any potential supplier of this service.

Applications

The oxygen cutting torch is the basic metal cutting tool found on most construction sites. Hand-held torches of this type are frequently the only cutting tools needed for major dismantling jobs, both above and below water, on steel and cast or wrought iron up to about 300mm (12in) thick, when a slow, labour intensive procedure is acceptable. Moreover, even the contractor employing more specialist techniques will often need to supplement these with a cheap, handy, method for cutting any difficult steel that is encountered and the basic oxy-acetylene kit is ideal for this purpose.

2.3.3 POWDER CUTTING

Powder cutting is a technique for boosting the combustion temperature of any oxy/fuel gas mixture and increasing its effectiveness through introducing a stream of fine powder, usually iron or iron-aluminium mixture, into the flame. The resulting temperature of 2800–3500°C, depending upon the mixture used, produces an effect similar to that of the thermic lance, melting the concrete and forming an iron silicate slag which flows away from the cutting area.

Because the flame must freely traverse the full depth of the cut, it must be started from a free edge or from a hole which has previously been formed through the member, by some other means.

There are two forms of equipment which can be used.

2.19 Dismantling a disused railway viaduct by oxygen cutting

2.20 Powder cutting non-ferrous metal

(a) THE POWDER TORCH

Powder cutting is a continuous process and can be carried out with quite simple hand-held equipment, consisting of an oxy-acetylene torch fitted for powder injection. In addition to the usual gas cylinders there would be a pressurised storage and dispenser vessel for the powder plus a cylinder of nitrogen for dispensing the powder. However, such elementary equipment is rarely used because it is too slow and the maintenance of successful cutting conditions too difficult for most operators, especially in damp conditions which impede the flowing properties of the powder.

More advanced powder cutting torches are available, which are designed to traverse the workface on automatic feed mechanisms, and these offer significant advantages because more of the operating variables can be controlled.

Torches will operate at any angle below the horizontal, including vertically downwards, and they produce narrow cuts with an excellent finish. A number of cutters can be set to work on the same workface and work 24 hours a day with only limited supervision and no operator fatigue. Furthermore, the cost of the consumables is comparatively low; for example, one recent job involved cutting heavily reinforced concrete 460mm (18in) thick, in which the powder cost worked out at $1/in, at a cutting rate of about 1½in/min, equivalent to approximately £25/m, at a rate of 40mm/min.

.21 Powder cutting a 750mm (30in) thick concrete bridge deck in 'alifornia, using a modified Union Carbide torch

pplications

'owder torches are suitable for use on thicknesses of concrete up o about 1m, and may be considered as an alternative technique o the thermic lance especially when an improved finish is equired, on almost any cutting duty other than for holes or very eep cuts.

b) THE POWDER LANCE

'he powder lance brings together some of the features of the owder cutting torch and the thermic lance. As with the thermic ance, it is a manually operated tool best suited to punching oles, and is in fact, almost identifical in appearance. It consists f a length of black iron pipe which is fitted into the powder ance handle, and a powdered-metal dispenser similar to that sed for the torch. Iron and aluminium powder are mixed with xygen in the lance handle, and burn at the open end of the tube. s the cutting action resulting from the oxygen and powder eaction takes place, the iron pipe is melted and consumed by the utting flame. These are the only consumables in the process, as t does not involve a fuel gas.

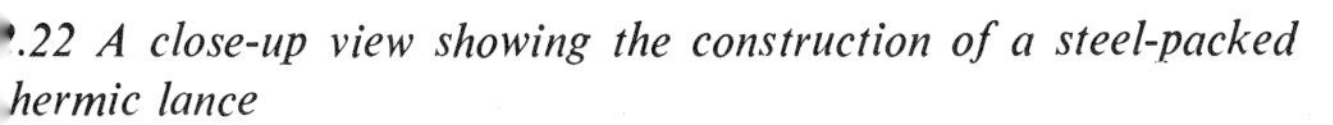

.22 A close-up view showing the construction of a steel-packed hermic lance

Applications

The flame produced by the powder lance is comparable in temperature and cutting efficiency to the best thermic lance. From the operator's point of view, it has the advantage over the thermic lance of weighing much less, since it is an empty tube which is being consumed. The powder lance is a useful tool for cutting heavily reinforced concrete over 460mm (18in) in thickness and has been successfully used to cut over 3·5m (12ft).

Neither the powder lance nor the torch have achieved the popularity among specialist contractors that has been attached to the simple packed thermic lance. This is probably because there is more equipment required and therefore more to go wrong in the powder processes, and the industry has been unable to attract or hold the trained personnel who could realize its potentially better performance.

2.3.4 THERMIC LANCING

The simplest of the devices to give a high temperature flame is commonly called the thermic lance, or "burning bar". The modern lance consists of a length of iron or mild steel pipe, usually about 3m (10ft) long, threaded at one end, which is packed with strands of iron wire. The threaded portion is screwed into the adaptor of the lance holder which incorporates an oxygen control valve. Oxygen at a pressure of about 0·7 N/mm² (100 psi) is passed down the tube to the open end, and is heated from any convenient source to start the process. Once ignited, the tube and wire are consumed in the oxygen, producing a temperature of approximately 2200°C depending upon the grade of iron used for the tube and filling. The only combustibles are oxygen and the lance itself. When the burning lance tip is applied to concrete, the silica will melt and combine with the iron to form an iron silicate slag, and ideally this should drain freely from the hole or cut. The slag will cool and solidify to a brittle mass which is then easy to remove.

2.23 Gas supply for a thermic lancing unit from 2000ft³ capacit gaseous oxygen cylinders

There are several different designs of packed iron lance being marketed which are more economical than the basic type. Thi improvement is achieved in one case by twisting the iron strand (Intrafix Inc) and in another by using an oval barrel (Weka lance). The intention behind these variations is to improve th oxygen distribution and hence the burning characteristics an economy. Alternatively, a higher output lance can be made b packing with a mixture of high carbon steel and magnesium alloy rods which will produce a temperature of about 4000°C (Thermischbranden Christiaanse). These lances are more expensive, but offer considerably reduced cutting time, and lance

2.24 A well-equipped thermic lance team using oxygen which ha been transported as a liquid in a specially constructed cryogeni tank

2.25 A thermic lance being used to modify a massive cast iron machine base. Some slag residue is visible beneath the base

consumption relative to the depth of hole. Using these, reinforced concrete 900mm thick can be pierced in 1½ minutes, consuming a length of lance equal to about four times the depth of hole.

Basically the advantages of the iron packed lance are the simplicity of the equipment needed, and the low cost of the lances. This is countered to some extent by the fact that they are heavy and awkward to hold when in use, expensive to transport, and are consumed rapidly.

The immediate equipment needed to carry out lancing is self-contained, but provision should be made on site either for handling and storing oxygen cylinders if the job is small, or for parking a liquid oxygen mobile tanker nearby and running supply hoses to the workface.

Applications

The device is most effective on heavily reinforced concrete because the steel aids combustion, thus increasing the speed of penetration. Holes up to 6m deep can easily be produced in a horizontal plane, the minimum hole diameter or width of cut, using a 19mm lance, is 60–75mm. Much smaller diameter lances are available, down to 6mm which may be used for making holes for passing cables or small bore pipes through concrete members and for severing reinforcement in cracked concrete. As a rough guide to the rate of progress, two 3m lengths of conventional iron lance should burn about 1m of 60mm diameter hole in 5 minutes.

Downward use of the lance, as in cutting floors, is limited by the fact that the slag cannot escape freely as in wall burning but must be blown clear by increasing the oxygen pressure. The effective operating depth of the lance in this application is between 600–760mm (24–30in).

The lance is also most efficient as a boring tool because the confinement provided by working in a narrow hole has the effect of concentrating the heat. For this reason the techniques for using the lance to make cuts are based on the idea of forming a line of holes along the intended breaking line, and then joining these together by other means, which are discussed below and in chapter 3.

In reinforced concrete, the position of the bars should be found by using a metal detector or cover-meter, and a series of holes then lanced at these positions. In mass concrete cutting, the spacing of the hole will be determined by the secondary breaking technique to be used to join the holes, and by the mass or size of the burden to be moved. Having formed this line of weakening holes there are three alternative routes to completing the break:

(a) join the holes together by "washing" away the interconnecting material between the holes by further lancing;

(b) use a concrete breaker to reduce the material remaining between the lanced holes, and then to complete the break with wedges, hydraulic bursters or rock jacks;

(c) in some situations it may not be necessary to carry out any further weakening if heavy duty rock jacks or some other means of loading can be applied so as to overcome the strength of the remaining material and break the mass.

The lance is equally suited to boring brickwork and building stone, and for cutting cast iron, wrought iron, and non-ferrous metals. It is particularly valuable for cutting old, fabricated iron structures where the laminations are usually separated by layers of rust which can render them nearly impossible to cut with an oxy-acetylene torch.

2.3.5 THERMIT POWDER CUTTING

Thermit powder is a proprietary material, available worldwide through the Thermit organization. It consists primarily of a mixture of iron oxide (commonly in the form of ground mill scale) and aluminium. When this mixture is ignited, through the

2.26a The first stage in cutting concrete with a thermic lance is to produce a line of individual holes

2.26b The lance is then used to "wash" away the material separating the holes, and thereby completing the cut

2.27–30 This sequence of pictures shows the final preparation, the firing, and the resulting severence of a rolled steel column by the application of Thermit powder

medium of an ignition powder placed in contact with the Thermit, an exothermic reaction of great intensity takes place, in which the iron oxide is reduced to iron, at a temperature of nearly 2400°C.

The heat from this reaction may be used to sever steel or cast-iron structural members. The method employed is to surround the member, at the position of the desired cut, with a temporary "crucible" which is filled with sufficient powder to provide the heat required. In practice, the temporary crucible is usually formed from a divided oil drum, lined with moulding sand. These are intended to be destroyed as the weakened structure collapses, but if there is any likelihood of the structure remaining in place, the crucible must be remotely pulled apart to allow the molten metal to drain away from the heated zone. If this is not done, there is a risk of leaving the newly formed iron mass to solidify around the structure, in fact strengthening instead of destroying it.

Applications

Thermit powder burning can only be applied to exposed metal structures. Historically, the use has been limited to the destruction of buildings in which a relatively large mass is supported on an exposed (or exposable) steel framework. Tedious and possibly hazardous dismantling of such a framework may be avoided by destroying certain selected members and allowing the collapsing weight to complete the demolition process.

Because Thermit powder is slow relative to explosives cutting, it is easier to programme and control a sequence of firing on a complex structure in order to achieve a predictable collapse or drop. Such a prediction is also helped by the fact that the process is vibrationless, thus making it possible to weaken the structure progressively to the point of collapse.

The charges are ignited by fuses or electrical firing devices from a safe remote location and this process is therefore recommended in the British Standard Code of Practice (CP 94:1971) *Demolition* as an alternative to explosives. This is of particular benefit when there is a risk that conventional demolition methods could place workpeople in danger if there was a premature collapse as might happen in an already unsafe building.

2.31 Hand-held flame spalling torch using an oxy-acetylene fuel mixture

2.32 The rocket-jet torch in use

2.33a,b,c The stone and concrete sculptures of Edward Monti produced with the Union Carbide oxy-kerosene rocket-jet torch

Although expert structural engineering knowledge is required to determine the placing and sequence for firing the charges, site preparations for this process are very simple and the material is completely safe and easy to handle even for unskilled labour, provided that commonsense precautions are observed.

Disadvantages of the process are that it does create a considerable fire hazard, the inevitable outcome of the emission of sparks and flames from the heat of the reaction, and also from the release, later, of streams of white hot molten slag as the crucibles are destroyed.

2.3.6 FLAME SPALLING

Flame or thermal spalling is a process for surface treatment of types of concrete and stone. The process differs from thermic lancing or powder cutting in that it is purely physical in its action, not involving any chemical reaction but relying for its effect on the fact that the mineral constituents of some concrete mixtures, and certain rocks, are very sensitive to thermal shock, and readily break up and detach themselves from the surface.

The ideal flame is one which is very hot, in the region of 1800–2300°C, and which has a high flame velocity. These characteristics will produce the desired thermal shock rather than a crude superficial heat which may be too damaging to the material being worked.

Straightforward surface spalling may be carried out with oxy-acetylene burners, but the process can be extended to shaping and channelling masses of material, by the use of an oxy-kerosene fuelled, rocket-jet burner. Both types of equipment will be reviewed.

(a) OXY-ACETYLENE BURNERS

This equipment is a development of the oxy-acetylene torches which are familiar in metal cutting; it is available in a variety of flame widths from 50mm (2in) to 750mm (30in). Small burners would be hand-held, but the large burners are mounted on a wheeled carriage, which is fitted with a hand-cranked traversing mechanism to give a uniform speed of coverage.

The gas consumption of the large machines is very high, and in order to use the machine safely it must be possible to move the required number of acetylene cylinders on to the site, while positioning them no further than 40m from the most distant point of working for the machine, in order that the pressure drop in the supply lines is not excessive. The 500mm burner will consume about 12,000 litres/hour of acetylene, and about 30,000 litres of oxygen. The acetylene is usually supplied from a bank of 18 or more cylinders to give the necessary flow rate (acetylene is only supplied in cylinders, in which it is stored in solution with acetone), but the oxygen is supplied either as a compressed gas in cylinders or as a liquid stored in a cryogenic insulated vessel. Some designs of burner incorporate the additional refinement of water cooled nozzles and the waste water is immediately ejected on to the heated zone, so as to increase the thermal shock.

(b) THE OXY-KEROSENE ROCKET-JET BURNER

The rocket jet burner differs from the conventional gas burning torch in that combustion takes place within a water cooled combustion chamber, and the expanding hot gases are discharged at supersonic velocity through a venturi nozzle (a convergent then divergent duct in which pressure is converted into kinetic energy by the acceleration through the waisted passage).

The fuel is a mixture of oxygen and kerosene, and the flame produced by the device has a temperature of about 2400°C and can achieve a velocity of 1830m/sec (6000ft/sec). The work surface is therefore not only heated more efficiently by this equipment, because of the improved heat transfer from the fast moving gases, but it is also subjected to a high velocity blast which dislodges and clears the spalled material. By careful direction of the blast, the surface can be formed into a desired shape.

Applications

Flame spalling is widely used throughout Europe for preparing concrete surfaces, both old and new, to receive new surface finishes, particularly epoxy-resin treatments. New concrete often has a surface layer of laitance, of poor abrasion resistance, and old concrete is sometimes contaminated with the remains of previous finishes, spilt chemicals (as in factory floors), oil and grease impregnation, or salts which have been put down to combat ice. Flame spalling will remove the damaged material, and will leave an inert dry surface ready to form the substrate for the coating material. (Note that all the loose material has to be cleared away by heavy wire brushing and subsequent vacuum cleaning before any coating can be applied.)

The technique may be employed to impart a decorative finish on several kinds of stone and concrete, and it is possible to vary the effects obtained by altering the gas mixture of the flame and hence its characteristics. An oxidizing flame, for example, usually produces, a clean, bright appearance, whereas a carbonizing flame will colour concrete in a variety of tones of black.

Stone and concrete shaping and channelling with the oxy-kerosene burner are usually carried out under open site conditions because the work is dangerously noisy, and it is difficult to control the smoke and flying debris. However, it is a most effective tool for rapidly producing complicated shapes, as is illustrated by the work of the American artist Edward Monti.

2.4 WATER-JET CUTTING

2.4.1 GENERAL NOTE: RANGE AND USES

Several manufacturers are currently producing pumps and equipment which will deliver a water jet of sufficient intensity to cut concrete, and most non-metallic building materials. Unlike the diamond or flame processes, the water jet, at the pressures available from the production pumps at present on the market, will not cut the reinforcing bars, or the individual fragments within the harder aggregates. Instead, it functions by eroding the cement matrix until the aggregate washes out.

It is important, in understanding the potential power of the water jet for cutting a material such as concrete, to remind ourselves of the nature of these substances. Concrete is not a homogenous mass, but a complex multiphase material which often contains as much as 10 per cent, by volume, of air and water. While it is technically feasible to apply sufficient pressure of water to reduce the whole mass to a slurry, this would be unduly wasteful of energy, and less rewarding than exploiting the weaknesses already present in the material.

There are probably three mechanisms by which erosion takes place:

(i) by impact loading of the cement matrix;

(ii) by cavitation on the sand and cement surface (the formation of low pressure zones which rapidly collapse);

(iii) by pressurizing the minute cracks in the cement paste and aggregate interface to the point at which the material will spall.

The efficiency with which these mechanisms translate the input power of the pump into the objective of achieving a satisfactory rate of cutting, is questionable. The minimum practical power requirements, with present jetting equipment, is about 120kW (160 bhp) which would give a cutting speed of around 700cm²/hr, while 224kW (300 bhp) is more usual. Various attempts are being made to raise the efficiency of the jet cutting action, by methods such as pulsed jets, the introduction of chemical additives to reduce the dispersion of the jet after it leaves the nozzle, and by the inclusion of abrasive compounds into the jet. However, most of the work on concrete is being done with conventional jetting equipment operating at moderate pressure and flow rates. Both pressure and flow rate are important factors

2.34 Starting the cut on a 250mm rc wall, using a water jet operating at a pressure of 83N/mm²

2.35 This water jet cut leaves the reinforcing steel in place and undamaged

in determining the cutting performance and the practicability of using hand-held water jet guns. Raising the flow rate has the effect of increasing the reactionary force at the nozzle, and this has to be compensated by the effort of the operator. At the same time a substantial flow rate is needed to give a sensible rate of cutting. In practice, this compromise is achieved by using pressures between 35 N/mm^2 and 120 N/mm^2 (5000 psi to 17,500 psi) and flow rates of between 35 litres/min, to 60 litres/min (9·2 US gallons/min to 15·8 US gallons/min).

Applications

Water jet cutting is a way of avoiding the fire hazard associated with the flame cutting methods, when a quiet, vibration-free cutting technique is necessary. In addition, it has in some circumstances the major advantage of leaving the reinforcing bars in the concrete completely undamaged, but exposed and cleaned.

A major disadvantage of this process is that the power requirements are considerable, as has been stated, and when this is

2.36 The high pressure water jet is an effective treatment for removing the kind of surface defects that are visible on this wall, and for leaving a finish of exposed aggregate

2.37 An operator using water-jetting equipment

considered in conjunction with the low cutting efficiency it can be seen to be a comparatively costly technique. Furthermore, the method is not without its hazards, as the debris which is dislodged, particularly the larger fragments of aggregate, is ejected from the cut with considerable force, requiring the operator to be suitably protected, and the area to be kept clear of personnel. Control of the waste water may also present a problem in that it is widely dispersed as a fine mist which may be airborne for some distance; a considerable quantity of waste water also requires disposal.

Considering all its characteristics, this process obviously has limited application. It is of especial use where the intention is to re-use the reinforcement by bonding into some new construction, or by welding a new system of reinforcement onto that already existing.

Also, being hand-held and a very flexible tool in use, the water-jet is frequently used for preparing random cracks in slabs or walls prior to filling with grout or sealant.

3 BREAKING: PRIMARY AND SECONDARY

The objective of cutting is the removal of material from a building member, whether merely to modify it or to remove it. Breaking, on the other hand, is not directly concerned with such a process, but is the operation by which the member itself is broken out.

If the overall objective of the work can be achieved by the use of breaking techniques alone, then they can be defined as *primary breaking*. However, primary cutting operations may have been carried out, and when they have been completed, the debris may be in the form of small particles of dust (say from drilling for small holes, or surface treatments), or it may be in the form of large sections of metal, concrete or brickwork. The process of rendering this debris into a form or size capable of sensible handling, where this is necessary, is that of *secondary breaking*. The techniques which are adopted for these operations have a quite different set of requirements from those of primary cutting They do not need to conform to the same standards of accuracy, or surface finish, but they obviously must meet the environmental standards. Some of the techniques of primary cutting will also be appropriate for secondary breaking, although they may be less accurately applied; heat/flame processes are sometimes used in this manner. For reasons of economy, however, the majority of this work is carried out with tools which are used only for this purpose, or for demolition.

The processes will be reviewed under the following headings, in each case with a discussion of their merits and disadvantages, and common applications:

3.1 Impact breaking;
3.2 Percussion tools, including hand-held and vehicle-mounted breakers;
3.3 Explosives, covering borehole charges, concussion charges and lay-on charges;
3.4 Hydraulic bursters;
3.5 Gas-expansion bursters (Cardox charges);
3.6 Hydraulic jacks;
3.7 The Nibbler.

Table 3.1 illustrates the range and use of the various methods of breaking.

3.1 IMPACT BREAKING

Impact breaking involves accelerating a comparatively large mass, which is brought into collision with the workface, dissipat-

Table 3.1 Breaking

Process	Applications	Preparation required	Degree of skill	Characteristics
Impact breaking (demolition ball)	total demolition on open site	nil	medium	fast and cheap, but limited to open site
Hand-held breaker	limited demolition	nil	low	widely available, and cheap, but labour-intensive, noisy and dusty
Vehicle-mounted breaker	limited demolition	nil	medium	powerful, but requires vehicle access
Borehole charges	controlled demolition, and fairly accurate breaking	study of structure, personnel control and pre-drilling holes	medium/high plus experience	all explosive methods can be cheaper and faster than alternatives for demolition, *if* circumstances are expertly assessed
Concussion charges	heavy demolition	ditto, but no drilling	medium/high plus experience	
Lay-on charges	heavy demolition and breaking	ditto, but no drilling	medium/high plus experience	
Hydraulic bursters	limited and fairly accurate breaking	pre-drilling holes for placement	low	quiet, vibrationless and controllable but slow and limited effect
Gas-expansion bursters	ditto	ditto	low	lower-skilled and less restrictive than explosives, but limited effect and variable results
Hydraulic jacks	heavier breaking of reinforced concrete	pre-forming holes for placement	low	useful for large mass or reinforced concrete, often in conjunction with thermic lancing
The Nibbler	concrete slab breaking	nil	medium	relatively quiet and vibration-free. Will dismantle slabs, walls or floors in a single operation. Requires heavy vehicle access

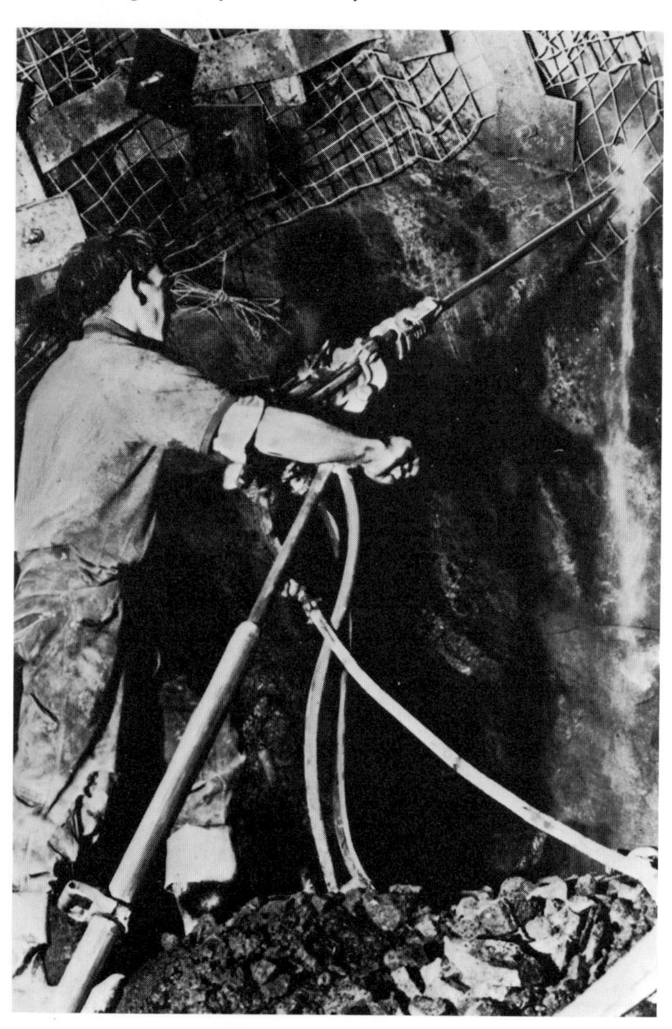

3.1 Pneumatic "feed" leg in use during rock drilling

3.2 Compressed air breakers have their place

ing its energy by reducing the workface to rubble. Small, manual impact tools, such as the sledge hammer, have been used for centuries. Their effectiveness was increased by mechanization on building sites, and pneumatic or steam hammers have been adopted for a variety of applications.

On a larger scale, the commonest demolition tool of this form is the steel ball, swung from the jib of a suitable crane or excavator. This consists of a specially made cast-steel ball which is impacted on to the structure by a motion imparted by the machine. The method used will vary according to the space available to manoeuvre the machine on site, and the plane and position of the intended target.

(i) The ball may be raised to its maximum height and allowed to drop freely, under its own weight;

(ii) or alternatively it may be swung in a horizontal arc by slewing the jib of the crane;

(iii) or alternatively it may be swung in a vertical arc in the same plane as the jib, by using a second cable, attached to the ball, to winch it towards the cab prior to release.

As an alternative to the steel demolition ball, a pusher arm is sometimes fitted to a heavy bulldozer, for use against low walls and columns.

Applications

These techniques are obviously most applicable to situations where destructive power is required, but where close control or accuracy of effect is not vital. Whether on a smaller scale within a building using hand-held tools, or on a larger scale using heavy plant, demolition is the principal application of impact breaking. Such techniques are unlikely to be useful in situations where close control of the process is essential, for example, where parts of the structure are to be retained, or where buildings are to remain in occupation. Even where the method and site appear suitable, the specifier must consider the transportation cost of any heavy plant involved, which ought to be recovered in the value of the work performed for the process to be economical.

3.2 PERCUSSION TOOLS

Percussion tools, as discussed earlier, differ in principle from impact breaking in that they rely upon a rapid succession of light blows, in contrast to the use of irregular massive blows. Although often inexpertly applied, with consequent disappointing results, percussion breaking can be a useful primary cutting method when used in the right circumstances. However, it has not been included in chapter 2 as its special significance to the users of the more sophisticated cutting methods lies in its application to secondary breaking.

The principle upon which these tools work in accomplishing breaking is to combine the action of the sledge-hammer and the spike to wedge the material apart. They are percussion tools and derive their performance from the total weight applied to balance the recoil; this will be the sum of the dead weight of the machine, plus the component of thrust added by the operator of hand-held tools.

From this stems one of the major disadvantages of hand-held percussion, and rotary percussion tools namely that they are best used underfoot, and are virtually useless when held above the horizontal. This can be overcome to some extent, with the rock drill, by using a pneumatic stand, and it does not apply to machine-mounted tools.

3.2.1 THE HAND-HELD BREAKER

The pneumatic concrete breaker (often erroneously called a drill or jack-hammer) is certainly the most familiar demolition tool in use today. There are also power sources for these tools other

3.3 A tractor mounted hydraulic percussion breaker of handy size for interior working

than air, such as self-contained petrol engines, electric motors and hydraulic systems, all of which have particular specialized advantages, as well as serious drawbacks vis-à-vis compressed air. An effective modern concrete breaker weighs around 23–27kg (50–60lb) and it is plainly impractical to expect an operator to compensate for the weight and recoil in anything other than downward use. For horizontal and upward breaking, the much lighter, but of course slower, single handed pneumatic pick should be used.

The effectiveness of hand-held breakers depends greatly upon the experience and ability of the operator, but is also dependant upon the age and strength of the work piece, and with concrete the aggregate it contains. In performing a common task, such as breaking up a layer of well-vibrated concrete with igneous rock aggregate, 150–300mm (6–12in) thick, overlying softer material, it should be possible to achieve a rate of 0·85m^3/hr (1yd^3/hr) but this could fall to around 0·6m^3/hr (0·75yd^3/hr). The wedging action of the specially shaped breaker point must always be used to move the portion to be broken off (the burden) towards a free edge. The correct choice of tool point is essential in order to achieve this effectively. There are a number of standard shapes and some proprietary steel points.

On very hard abrasive materials, and on mass concrete, these breaker points may not be effective, and an alternative method of using the pneumatic breaker may be adopted using the old stone-mason's practice for stone splitting. The operator will use a rotary percussion rock-drill to drill holes into which will be fitted split steel linings (called feathers), and a wedge will then be driven into the centre, using the breaker to provide the driving force. Holes should be drilled about 48mm (1$\frac{7}{8}$in) in diameter, to a depth suited to the length of the feathers available (normally 300–450mm long), and spaced according to the material being broken, so as to give a clean fracture along a line, rather than breaking out individual fragments. Three holes are usually drilled about 380–500mm back from the free edge. The two outer holes are then carefully wedged until the practised operator can tell by the "ring" of the steel that the material is in a state of maximum tension. The central hole is then wedged until the material cracks cleanly along the fault line.

It is important to note that the practice of using the machine and

3.4 The Montabert hydraulic hammer fitted to the digging arm of a Hymac vehicle

its steel (or point) as a lever to dislodge the material is not only unnecessary, but is strongly discouraged by the equipment manufacturers.

Applications

These tools are very useful for general breaking work, particularly of mass concrete and brickwork. They are widely available, and are comparatively cheap in terms of capital cost. They are capable of use by relatively unskilled labour, although as with most processes, skill and experience will assist in obtaining better performance; in the right hands, surprisingly good output can be achieved. The results will inevitably be crude, and for this reason the tools are most applicable to demolition works, rather than for alteration.

Noise and vibration are the two disadvantages usually associated with the techniques, and their effects must be considered when embarking upon a major session of this work. Vibration is not sensibly capable of elimination but there are ways of reducing the problem of noise, both for the operators and the public.

Such an improvement can be brought about by specifying that mufflers should be fitted to normal pneumatic breakers, or noise-reduced versions of the machine used, and that muted steels (points) should be fitted and if necessary all compressors should be of the fully silenced variety. For the operator's safety, it is recommended that ear-plugs, or ear-muffs should be worn whenever using this type of machine. Besides the avoidance of interruption to the works caused by injunctions from the local authority (see p. 48), there is the improvement in operator performance arising from lower exposure to the tiring effects of continuous noise. Persons controlling site works should be aware of the safety legislation on noise control which is in force in the country in which they are working.

3.2.2 VEHICLE-MOUNTED BREAKERS

There are several versions of percussion breaker which can be attached to the bucket arm or other suitable attachment point of tractors and digging machinery. They are invariably hydraulically operated from the power take-off on the vehicle hydraulic power system. Their action is similar to the driven wedge principle of the hand-held breaker, but the performance is greatly increased by the additional weight and power available. As an alternative to the sharp pointed steel, a blunt ended tool can be fitted which is more suitable for breaking up large lumps that have already been detached from the main mass.

Applications

These machines have a similar role to hand-held breakers, in principle, but with the emphasis on thicker materials, which would take a long time to break using hand tools.

The advantages of using tractor mounted breakers are the greatly improved output and the ability to work on vertical faces and floors at considerable height above the working floor level. This capability will obviously vary considerably from one design of tractor to another, depending upon its stability and the reach of the arm. The major disadvantage is obviously the requirement for machine access, and they cannot normally operate at heights above the reach of the machine standing at ground level.

3.3 EXPLOSIVES

As many demolition operations and some cutting can be carried out by blasting with explosives, without particularly onerous restrictions, their use deserves the designer's consideration. The technique can be applied to structures built with virtually any material such as stone, concrete, brick, steel and wood. Speed, low cost and in some special instances *increased* safety are the attractive features of the method. Increased safety may be obtained because the destructive force is remotely initiated and personnel can be kept well clear of the effects of the work.

It is recommended that in the event of explosives being specified as the method to be employed, this should entail the appointment of a competent and highly experienced specialist contractor. That being so, such a contractor will be knowledgeable on the subject of safe handling and detonation of the charges. Competent contractors will always carry out a site survey before committing themselves to carrying out the work, and in the course of the survey they will usually complete a check list (see chapter 7.2).

In some circumstances the use of electrically detonated charges may carry the risk of premature explosion due to the effects of stray currents, or electromagnetic radiation. The likelihood of such incidents is very remote, but it is worth pointing out to the contractor if anyone is aware of hidden current-carrying cables, or the existence of nearby radio transmitters, radar or fixed navigational transmitters.

The following aspects of explosive work need to be carefully considered.

(a) Site control and safety during explosive cutting

The explosives should be under the control of an operator who is licenced by the police to possess and handle them. The shot-firer will have responsibility for all the precautions before firing the blast, ie:

(i) he must make sure that a clear warning or signal is given;
(ii) he must make sure that all approaches are effectively guarded so as to prevent access by persons while the shots are being fired;
(iii) he must make sure that all persons, including himself, have taken adequate cover.

(b) Vibration from blasting

The designer's prime concern will be with the effectiveness of explosives in achieving his objective, and the avoidance of unpleasant side effects, the obvious ones being the danger of flying debris and the effects of vibration. The former may be guarded against by fairly simple precautions, but the latter remains as a factor to be considered. It is worth remembering at this stage of the forward planning that adverse reaction from the public may be based upon subjective fears regarding safety when vibrations are still well within the safety limits of the structure.

(c) Ground vibrations

The parameters which determine the effect of ground vibrations on structures are the amplitude (displacement) and the velocity (or acceleration) of the ground movement. Recent work tends to support the importance of velocity in assessing the risk of damage, and the peak particle velocity of vibration is now accepted as the best criterion to use. Velocity takes into account both the frequency and the amplitude (expressed in the relationship $V=2\pi\ fa$, where V=velocity, f=frequency in Hz and a=amplitude in mm) and gives an indication of the possible hazard of the movement, as well as a guide to its nuisance value. The United States Bureau of Mines recommends that vibration levels should be kept below a peak particle velocity of 5·1cm (2in) per sec. British practice has centred around the work of the late Dr G. Morris of ICI Ltd, who suggested various maximum levels of displacement depending on the type of property.

(d) Air blast

The effects of minor air blast are to cause loose doors and windows to rattle, and arouse people's concern. These effects are often incorrectly attributed to ground movement. Probably the fact that these events are accompanied by a sudden noise increases apprehension, although the actual manifestation is no more than would occur naturally on a windy day.

3.5 The line of rock drilled holes have been prepared to provide a break line for the subsequent firing of explosive charges

In practice, the shock wave diverging from the area of the blast rapidly degenerates into sound waves only. Values of air blast are normally quoted as an over-pressure in kg/cm^2, or to a log scale in decibels.

Windows may be broken with over-pressures in the range of 0·05–0·14kg/cm^2 (0·75–2·0lb/in^2), and as little as 0·007kg/cm^2 (0·1lb/in^2) may be sufficient to crack a badly mounted pane; 0·003kg/cm^2 (0·05lb/in^2) is sufficient to make a loose sash window rattle.

As with all breaking techniques, an element of unpredictability exists, and the contractor will apply the experience gained from the first shot firing to adjust the procedure. His aim will be to leave a good free face at the end of each firing. Then by clearing away the debris and carefully placing successive holes, the risk of flying debris can be greatly reduced.

3.6 The explosives used to make this shaft breakthrough have left the reinforcement undamaged whilst stripping the concrete

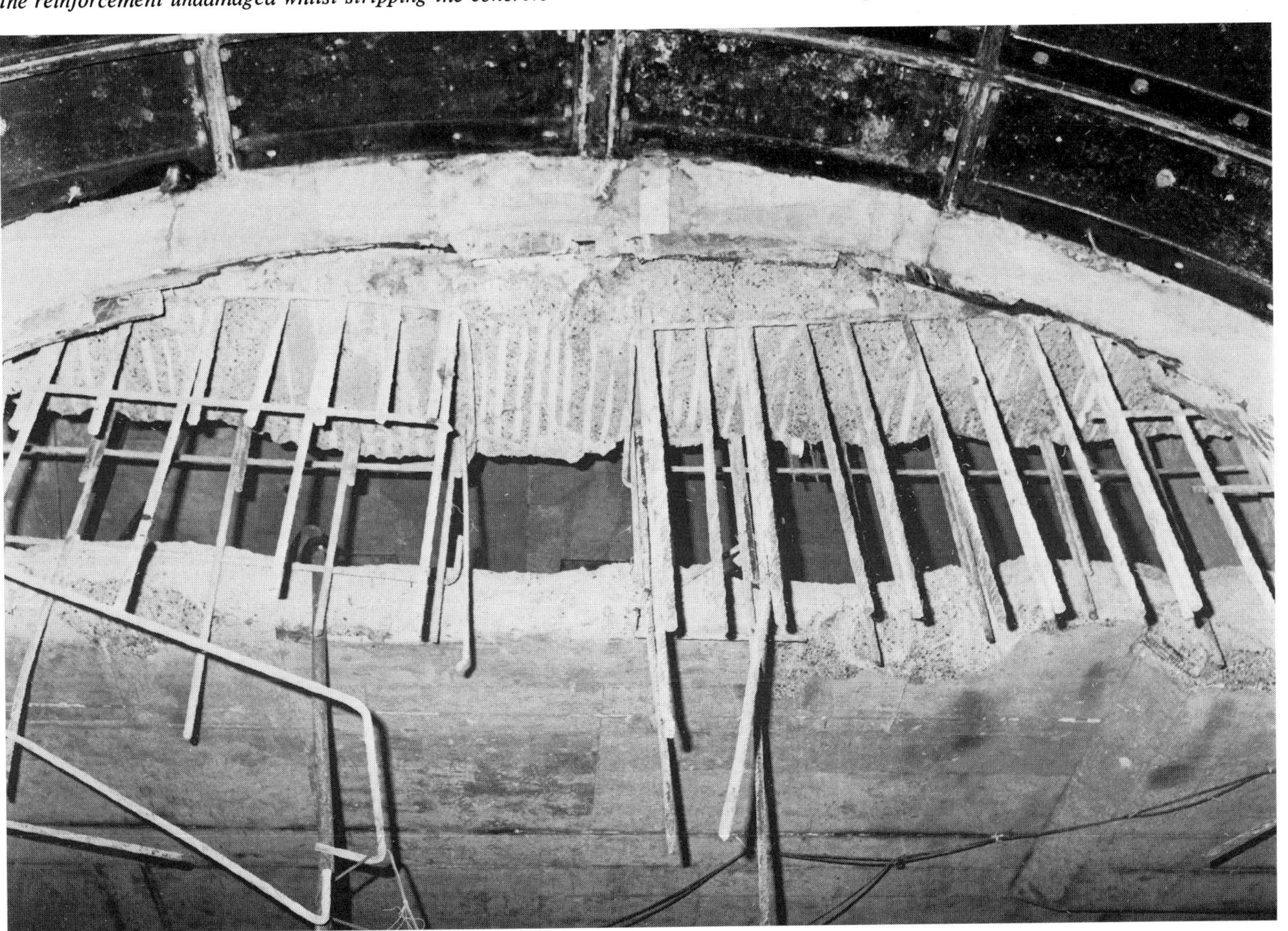

Timber baulks, or sandbags, laid over tarpaulins, are frequently used for covering the charge, although both take a long time to place, and the sandbags quickly disintegrate. Fine mesh steel wire net may sometimes be used instead of the tarpaulin because of its longer life, but it is not satisfactory on its own for containing small fragments.

The types of explosive commonly used are:

Single component explosives	**Form**
Nitroglycerine mixtures	powder or gelatine
TNT/Ammonium nitrate	gelatine
Blasting agents/slurry explosives	slurry
Black blasting powder	powder
Two component explosives (both safe until mixed)	
Astrolite*	two liquids

There are three ways in which explosives can be applied, and each will be reviewed, with its applications, under the following headings:

3.3.1 Borehole charges. The explosive is placed in holes drilled at strategic points in the structure.

3.3.2 Concussion charges. Here the explosive is placed as a bulk charge in the centre of the structure to be broken up.

3.3.3 Lay-on charges. With this form of charge, the explosive is placed in contact with the object to be reduced, and then confined by means of clay or sandbags. Shaped and linear charges fall into this category.

3.3.1 BOREHOLE CHARGES

This is probably the commonest form of use of explosives in the construction/demolition industry, in which the explosive charge is placed in holes drilled at strategic points in the structure. When competently applied they enable reasonably accurate and

* Trade mark of the Explosives Corporation of America.

selective work to be undertaken. Most elements of a building can be tackled, whether simultaneously for widespread demolition, or individually for local clearance or alteration.

As this work is generally undertaken inside existing buildings, it is important to prevent the projection of debris. This is achieved in two ways, both by using small closely spaced charges, and by ensuring that they are properly covered. The usual approach is to blast a free face away from the main body, the work progressing in stages depending on the size of the burden that can be blown safely at one time. Vertical holes are normally preferred to horizontal ones as any debris that is projected will travel more or less horizontally before falling to the floor. This is in contrast to horizontal holes which might have the effect of launching some of the debris into a far-travelling trajectory.

Holes are usually drilled at a spacing equal to the burden (distance to the free edge), and it is recommended that this distance should be about 450mm (18in). The depth of the holes should not be much more than one third of the depth of the material, say about 600–900mm (2–3ft) in an 1800mm (6ft) deep block. Each hole would be charged with between 55 and 115gm (2 and 4oz) of a medium to high power gelatinous explosive, and two or three holes fired together using instantaneous electric detonators.

Applications

(*a*) *Concrete floors and roofs*

Concrete floors and roofs can be more speedily removed by the use of explosives than by almost any other method when circumstances allow. The technique adopted is to drill holes about 150mm (6in) deep obliquely into the slab at 300mm (12in) centres, and then to charge each with a suitable explosive; as many as 30–40 holes may be fired together. Afterwards a breaker is used to complete the work, and it is important to note that these methods will not cut any reinforcement that is present, so this must also be tackled separately.

(*b*) *Brick and concrete walls, piers and pillars*

Although plaster charges can be used for removing small walls, the borehole method is preferable for thick walls, buttresses, and retaining walls. Two or three rows of holes are drilled into the vertical face of the wall, inclined downwards at an angle of 45°, and of such a depth that the charges are situated centrally within the mass. The holes in one row are staggered in relation to those in the rows above and below them, and the bottom row should be at ground level to achieve a good shearing action.

Well-constructed brickwork can be remarkably tenacious, and to ensure the proper collapse of the wall at least two rows of holes are always used. For thicker walls (900mm (3ft) or over), three rows are preferable.

Walls or piers which are broad in proportion to their length aer very stable, but if three rows of shot holes are used, the aim should be for the structure to drop so much that it will become unstable and topple. Walls must be free at the ends, otherwise the charges may simply blow out the base, leaving the wall bridging between the supports. The weight of charge used by the contractor will vary widely according to the nature and thickness of the wall, and to some extent, the weight above the level of the blast.

Very thick walls and piers may have rubble cores, and it is important for this to be identified when drilling the holes. Care should then be taken by the contractor to place the charges only in the solid material.

This type of explosive cutting will probably leave substantial pieces of debris unbroken, because much of it will have only been subjected to the force of their fall. Some form of secondary breaking, possibly again employing explosives but in the form of plaster charges, or alternatively pneumatic breakers, will be required.

As an alternative to felling a wall in one piece (which may be unacceptable if the designer is concerned about the resulting floor loading), walls may be disintegrated in position by placing occasional holes all over the surface, and firing these charges at the same time as those in the base. If the member has a flat top, and is not too high in relation to its width, good results can be obtained by drilling one or more rows of holes vertically the full depth of the wall. This gives excellent fragmentation but means that a considerable depth of drilling is required.

3.3.2 CONCUSSION CHARGES

In this case, the explosive is placed for firing in the form of a bulk charge in the centre of the structure, or more rarely at a number of points within the structure.

Applications

Because their effect is not immediately confined, concussion charges are not capable of sufficiently close control to be used in situations where parts of the structure are to remain intact, or where other activities or buildings are in close proximity. They are, however, of great value in breaking up retorts, tanks, hoppers, and semi-enclosed vessels of the kind commonly found in chemical and process works. Generally these structures are made of cast iron, boiler plate, or welded construction, but the method has been found to be equally effective on thin reinforced concrete containers. The vessel to be broken up is filled with water, and the charge suspended in the centre. The energy released by the explosion is transmitted through the water to the surrounding structure, but as water is virtually incompressible it ceases to exert any force once the pressure is released at any point. Thus, while the water acts as a good transmitter of the shock waves and therefore only light charges are required, once the structure is broken the danger of flying debris is reduced.

3.3.3 LAY-ON CHARGES

In its simplest form the explosive is one which is placed in contact with the object to be reduced, and then confined by means of clay or sandbags applied around it. While this crude use of explosives may be acceptable in some situations as there is little preparatory work, more precise objectives may be achieved by particular forms of "lay-on" plaster charge, called either "shaped" or "linear" charges.

In the former case, the explosive is placed in a specially designed container which forms a hollowed-out wedge or cone shape in the base of the charge and serves to direct the force of the explosion so as to achieve closely controlled results (the Munroe effect). The latter description is applied to a technique for attaching plaster charges, made up from high velocity explosive in the form of a continuous strip. The successful application of shaped or linear charges requires considerably more knowledge of the behaviour of the materials and explosives than the more straightforward applications. In the case of shaped charges, these have to be engineered specifically for the task, so that most of the preparation takes place in the factory and little of the detailed work can be seen on site.

Applications

(*a*) *Steel plate and sections*

Girders are usually cut by placing two charges, one on each side, but staggered by an inch or two, so that a shearing effect is produced. The two charges must be initiated simultaneously. This method is useful in difficult situations where an oxy-acetylene torch cannot be used, or when the structure is likely to collapse when the member is cut.

3.7 Hydraulic breaker with a compressed air-powered hydraulic pump, which converts air pressure at 1N/mm² into the required hydraulic pressure of 125N/mm²

(b) Masonry, brickwork and concrete

Plaster charges may be effective on walls up to 350mm (14in) thick, but over this thickness the borehole method is preferable. In removing a wall, the charges are generally placed at about 300mm centres, and at about the same distance from the ground level. Masonry requires the use of heavier charges than brickwork, and still heavier charges are needed for reinforced concrete; when tackling reinforced concrete, the usual practice is to use the explosive to blast the concrete off the steel reinforcement, which then may be cut up separately.

For the formation of openings, or avoidance of damage to adjacent areas which are to remain, the contractor will usually drill a series of weakening holes to produce a fault line along which to break. This may be supplemented by the use of shaped charges which can be directed accurately.

3.4 HYDRAULIC BURSTERS

Reference has been made earlier to the stonemason's technique of driving wedges into pre-drilled holes lined with split steel tubes (or "feathers") in order to detach blocks of stone of accurately determined size. The method has been updated in the form of hydraulic bursters. The concept of lining the holes, as a means of spreading the expanding force so that it stresses the mass of material rather than merely causing local crushing, is used in all these devices.

The expansion force is now obtained by a mechanism for applying hydraulic pressure, rather than the earlier driven wedge. Two forms of such mechanism are commonly used, the plunger burster and the wedge burster. In the former, hydraulic pressure of up to 125N/mm² is applied to a system of plungers which are forced out from a central cylindrical core. This device expands in one direction only, and therefore has to be carefully orientated both in relation to the lining, and to the direction in which breaking is to occur. The second design of burster employs hydraulic power to retract a steel wedge between tapered steel liners. In this device these "feathers" are an integral part of the assembly. In either case, the bursters are often used in multiple arrangements, fed from a common hydraulic power supply.

Applications

Hydraulic bursters can be employed both in primary cutting procedures, and for secondary breaking. They can only be used when some movement of the burden can be tolerated, which obviously limits their primary uses.

Large pieces of debris resulting from primary cutting operations are often subject to secondary breaking by bursters, particularly where noise and vibration would be a problem.

Hydraulic bursters have distinct advantages over many other breaking methods, and indeed over some primary cutting processes, in that they are almost free of noise and vibration. Although rarely capable of breaking reinforcement, they are quite often used in reinforced concrete where other appropriate cutting methods have removed the reinforcement and possibly also effected some additional weakening of the concrete before the bursters are applied. This will obviously be most applicable to larger masses of concrete, or large free-standing members.

3.5 GAS EXPANSION BURSTERS (CARDOX CHARGES)

Gas expansion bursters use the force generated by the expansion of a highly compressed gas, being suddenly released at a strategic position within the material, to over-stress the mass beyond breaking point. The highly compressed gas is liquified carbon dioxide, contained within a steel shell; the sudden release of destructive energy is achieved by first electrically energizing a chemical charge which causes a rapid build up of pressure until a bursting disc, restraining the CO_2, ruptures and releases the gas pressure. The storage pressure of the CO_2 gas prior to energizing the charge is about 20·6N/mm² (3000 psi). The characteristic of the released force is to produce a "heaving" thrust, rather than a sharp explosion.

The charges are fitted with retaining pawls which are forced into the walls of the borehole by the force of the escaping gas, so as to prevent the shell from being propelled out of the hole. However, these cannot be relied upon, and it is essential to apply some additional restraint, in the form of heavy weights and sandbags. Cardox charges are readily available from the suppliers in the UK and the USA, and in many other parts of the world, or alternatively the empty shells may be purchased and filled on site. The shells are obtainable in a range of pressures and physical dimensions to suit most applications. The bursting pressure lies between 123N/mm² (17,500 psi) and 270N/mm² (39,000 psi), and these may be applied with bursters which are available in a range of lengths between 685mm (27in) and 1244mm (49in) long. The method of applying the charges is very similar to that used for borehole explosive charges. The boreholes are prepared by rock drilling to a suitable depth in the mass material, taking care to ensure that there is sufficient thickness beyond the bottom of the hole to contain the pressure and direct the energy to removing the burden, instead of merely blowing out the end of the hole. The holes need to be fairly accurately drilled to a diameter no more than 3mm (⅛in) greater than the diameter of the shell to be used. When the shell is placed in the hole, it is advantageous to fill the remaining space with water, if this is possible, as this has the effect of transmitting the force to the material to be broken, without any loss.

In the UK the charges may be handled by anyone without a police licence, and are therefore not as restricted in their use as are explosives. Some experience is necessary in order to obtain the best results from their application, and in the wrong hands they can be extremely dangerous. For these reasons, the specifier should ensure that the contractor has the required experience, and that the manufacturer's instructions are followed.

Site control procedures similar to those applicable to the use of explosives should be adopted to safeguard site personnel or the public who might be present in the firing area. The same precautions as are necessary for small explosive charges must be applied to contain any flying debris, although additional safety is obtained from the fact that the gas which is released is both inert (it will not support combustion), and extremely cold.

Applications

Cardox charges are normally applicable to primary operations only in relatively open situations, and for groundworks. They are useful tools for secondary breaking, particularly in more massive members or large debris.

Their avoidance of some of the restrictions placed on explosives, and their less violent effect, can be advantageous in fairly precise breaking work. However as their action is dependent on a build up of gas pressure in the workpiece, it is normally only in thick members that this can be harnessed. In these terms, they may generally be regarded as an alternative to explosives.

3.6 HYDRAULIC JACKS

Hydraulic jacks advance the concept of over-stressing material in tension to breaking point, by enabling forces as high as 350 tonnes to be exerted at a single location within the mass of the material.

3.8 An operator preparing to fire a Cardox charge. The 1m thick heavily reinforced concrete base has been cracked, and a piece detached by a previous single charge

The jacks used for this purpose are similar to those used for industrial and vehicle repair applications, except that they are very ruggedly constructed; this is to enable them to cope with the rough conditions on building sites, and to withstand off-centre loading which often occurs. Also they are designed to be supplied by a hydraulic pump which is remotely sited, through a system of control valves which regulate the applied load and the retraction of the ram.

Jacks are generally used in groups of three, or multiples of three, positioned in holes in the workpiece which are spaced apart to suit the thickness and strength of the material and the direction of the desired break. As an example a typical arrangement for breaking a section of 750mm (2ft 6in) thick reinforced concrete wall would require three jacks. These would be placed in holes approximately 150mm (6in) diameter × 250mm (10in) deep, cut by flame methods or diamond drilling, and spaced approximately 1m (3ft) apart. The jacks used for a task of this magnitude would normally have capacities of about 100 tonnes each. The amount of wall broken out at each stage would depend on the tensile strength of the reinforcement present; if only light reinforcement is present, the weight of burden broken obviously relates to the jack capacities.

Applications

Hydraulic jacks are mainly used in the dismantling of heavily reinforced structures when primary cutting techniques have first eliminated reinforcement and/or weakened the material along the intended break line. Used in this manner they are extremely effective, and minimize the environmental disturbance, and effect on the residual structure. While they are virtually silent and vibration-free in operation, jacks have the disadvantages of requiring large holes to be pre-cut in the material in order to accommodate them, and a considerable amount of movement is required in order to break substantial reinforcement in reinforced concrete.

3.7 THE NIBBLER

The Hymac Nibbler is an item of tractor mounted breaking equipment for concrete breaking which works on a novel principle, and offers a quieter, faster alternative to percussion methods. It is a device for applying a bending load to a concrete member so that it snaps. Jointly developed by Hymac Ltd and the Building Research Establishment of the UK, the Nibbler

3.9 Hymac Nibbler breaking up a reinforced concrete road slab

has proved to be effective in breaking reinforced concrete up to 380mm (15in) thick, containing 20mm ($\frac{3}{4}$in) reinforcing bars, without vibration and with comparatively little noise.
The breaking output is high, as much as $100m^2$/hr ($120yd^2$/hr) with the additional facility of being able to break and then lift pieces up to $1\frac{1}{2}$ tonnes for instant removal from the workface. The various breaking actions which can be achieved are shown in 3.10 in the series of sketches.

Applications
The most common application of the Nibbler is the breaking and removal of concrete oversite slabs. However, the machine is extremely versatile, and when fitted to a Hymac hydraulic excavator as a front-end tool it is capable of breaking vertical walls and suspended slabs; obviously it is not normally practicable for it to operate on suspended floors, due to its weight.

3.10 The arrangement of the Nibbler jaws to provide for both front and back clamping action
3.11 The application of the Nibbler to oversite concrete breaking

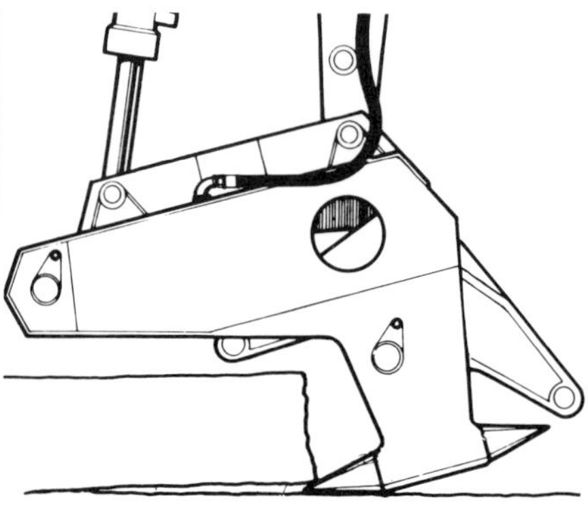

(i)
The parallelogram configuration and the full bore power of the dipper ram give immense penetrating power, pushing the tooth easily under the concrete. The Nibbler remains horizontal with minimum manipulation.

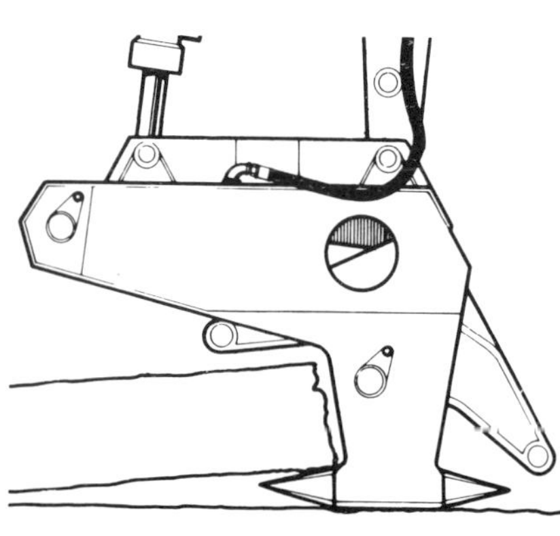

(ii)
When the concrete is resting against the root of the tooth, the Nibbler is rocked back, using the bucket ram. This prises up and applies a bending moment to the concrete.

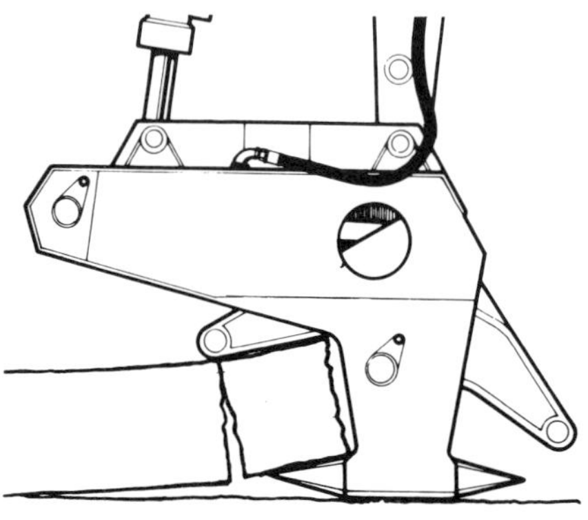

(iii)
The Nibbler jaws are closed down by the clamping ram, breaking the concrete between the root of the tooth and the rocker beam anvil.

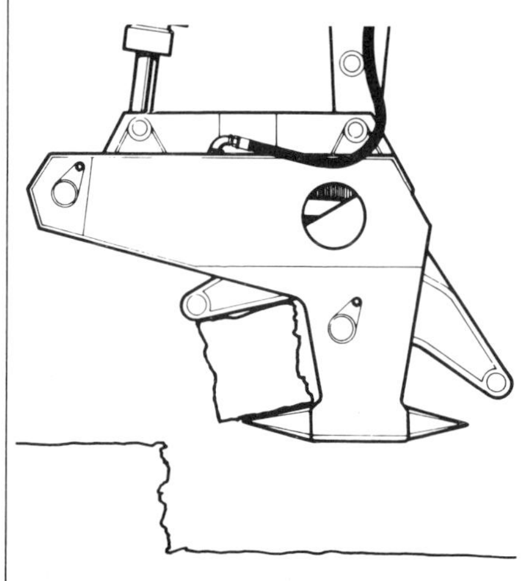

(iv)
With the rocker beam clamping action continued, the broken concrete is gripped and moved out of the way, giving clear access to the next bite.

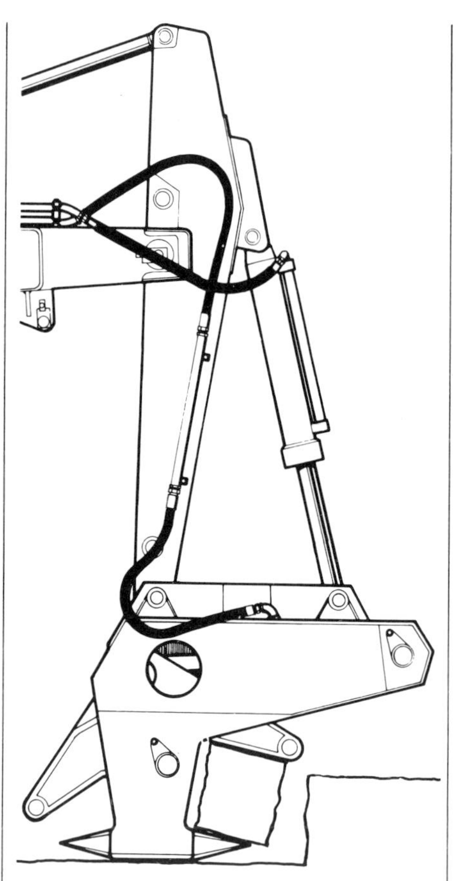

(v)
For smaller jobs, or thinner concrete, the Nibbler can be mounted to act forward without moving the bucket ram. In this configuration, whilst not as powerful, the Nibbler can be used to break vertical walls and first floor level ceilings.

4 SITE FACILITIES AND SUPPORT SERVICES

Specialist cutting and breaking is generally carried out by very small teams of operators, and the specialist subcontractor must endeavour to keep the scope of his work and responsibility to a minimum. The successful completion of anything but the simplest job may therefore be dependent upon the attendance of others, plus the provision of some services and facilities. These items cover physical access and handling, power and water supplies and disposal, and the removal and disposal of the debris from the operations.

The specialist contractor's advice should be sought to ensure that the facilities provided are adequate, and also to check that they do not limit the ability to perform the cutting task. At the end of the chapter, chart **4.5** illustrates the sequence of operations and normal contractual responsibilities, and chart **4.6** shows which requirements arise with each of the common processes.

4.1 SITE ACCESS AND LOADING

While limitations on access to the site, and upon the loads that can safely be imposed on various levels and surfaces, have quite obvious implications for the types of plant that can be employed, and consequent influence on the choice of method, they also have less obvious effects.

The most important of these effects to be borne in mind by the work organizer is that arising from the dimensions and state of any redundant material which needs to be removed from the site, ie the debris removed from the structure. This is a "circular problem", in that there is no clear starting point for identifying the constraints which will ultimately dictate the approach to be used.

If the restrictions are examined sequentially, they will yield a consecutive series of handling problems, although they may not all arise in every case.

It is generally the most cost-effective use of specialized cutting techniques when the secondary breaking is kept to a minimum. However, this may mean dropping a heavy mass of material on to floors which have to remain in place, then moving the mass through a tortuous route across the site, and finally being left with a problem piece of material which no disposal contractor will accept. Even when a sensible amount of secondary breaking has been done, the debris may still be in pieces too large to be easily handled by bucket loaders, or prove to be too damaging to drop into skips.

Finally there is the need to question the selected cutting contractor about the specific type of plant or equipment that he proposes to use on the site. There are so many variations, even within basically similar methods, of the shape, size, and weight of the gear that may be used, that it is unwise to make more than very broad, generalized assumptions on the implications of any particular method. Thermic lancing, for example, requires a substantial supply of oxygen; this may be dispensed from banks of standard industrial gas cylinders, or from smaller numbers of very large cylinders (usually containing 2000ft^3 of free gas), or from a liquid oxygen storage tanker. Each of these will have advantages and disadvantages according to the constraints of any particular set of site circumstances.

A great deal of the equipment that has been referred to relies upon energy sources that are remote from the workface, whether gas storage, compressors, hydraulic pumps, or electric generators. In nearly all cases there will be working limitations on the

4.1 Site access and loading operations

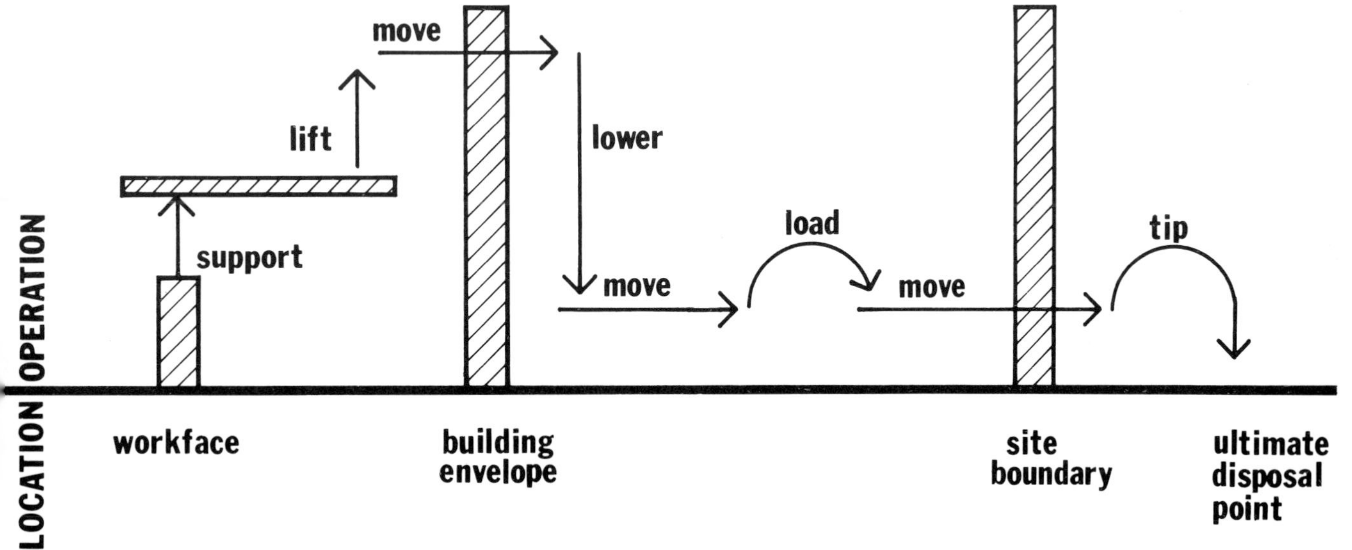

4.2 Dismantling an rc frame structure using a powder torch. Note the careful propping for the severed beam

lengths of interconnecting cables or hoses which will mean that parking space must be planned for the appropriate piece of plant within the operational area.

4.2 MAIN CONTRACTOR'S ATTENDANCE AND FACILITIES

The attendance may start with the need to provide labour to off-load the equipment and to place it into position for work; the need for the main contractor's craneage, particularly in tall buildings, is a frequent source of friction and delay to many site operations. Temporary propping, and subsequent support for any redundant material, will also feature among the main contractor's responsibilities, along with that of providing staging and safe access to the workface. If the particular situation and operations require the erection of screening as a protection against the dust, smoke, fumes, or water, then this should be in place before the specialists arrive on site, so as to avoid costly standing time.

Protection against the effects of the waste flushing water from the diamond tool processes is best provided by control methods applied close to the source, as has been described in the section on processes (pp. 12–19). The need for this specialized equipment, plus the fact that each change of cutting tool position will necessitate some alteration in the water collection arrangement, indicate that this aspect is one which is best dealt with by the cutting contractor himself.

4.3 MAINS AND OTHER SUPPLY SERVICES

Whenever possible, it is preferable for reasons of reduced operating costs to make use of mains supplies of water and electricity. A standard domestic water supply (around 3 gallons/min through a ½in diameter pipe) will suffice for most applications, subject of course to the possible need to boost the pressure by means of a pump, so as to maintain the supply at a height above the working head pressure available.

Electricity is a convenient power source for workface lighting, waste water pumping, and fume extraction fans. An electrical supply is also sometimes needed to power small cutting machines such as saws or drills, or to run ancillary equipment.

Single phase line voltage (230/240 volts at 50 Hz in the UK) will power most of these items, either directly, or usually for increased safety, through an isolating transformer. Larger pieces of equipment, such as electro-hydraulic power packs and solid state frequency changers (these change 50 or 60 Hz supply to 400 Hz without any moving parts), for example, will require a substantial 3 phase power supply.

On many large sites where extensive works are in progress, there is often a plentiful supply of compressed air, and it is common practice for the cutting contractor to draw from this supply if he is employing pneumatic tools. However, this is not altogether a satisfactory arrangement, in spite of the superficial advantages, because a number of problems can arise. Pneumatically powered cutting equipment is frequently under-powered for the job it is expected to do, and the slightest reduction in supply-efficiency can render it practically useless. For this reason, pressure drops and leakage losses in the air lines must be minimized, and the compressors themselves well maintained, so that the rated output is in fact achieved. Other pneumatic equipment found on construction sites is rarely as sensitive in this respect, and the needs of the cutting contractor may therefore not be appreciated. The resulting poor performance can prove costly to the specialist, the main contractor, and ultimately the client, depending upon the contractual conditions. In addition, because the air demand

4.3 Drilling exterior fixing holes with a diamond drilling rig securely braced with scaffolding

of some tools is very high, up to 5m³/min (180cfm), there may be considerable "hidden" costs in running the compressors, which will be difficult to identify for their correct allocation against the cutting expenses.

4.4 DISPOSAL OF DEBRIS

Under the normal contractual arrangements agreed in the UK, the main contractor is almost always left with the responsibility for removing and disposing of the redundant structure or debris. While the disposal of core material from drilled holes or fragments and slurry from surface treatments should be fairly easy, it may prove difficult to find an ultimate resting place for some large masses of concrete or other material; this may entail searching for another specialist with the equipment to lift and transport such loads. These problems, and the costs associated with their solutions, must be balanced against savings which may have been made from a reduction in the amount of secondary breaking. In other words, the total costs of achieving the constructional objective embrace many more factors than simply the basic cost of the original cutting operation.

4.4 A massive section of concrete wall detached by primary cutting by thermic lance. The contractor may have the choice, in a similar situation, of carrying out secondary breaking or of disposing of the redundant section in one piece

4.5 Facilities and support functions shown in chronological sequence with normal contractual responsibilities

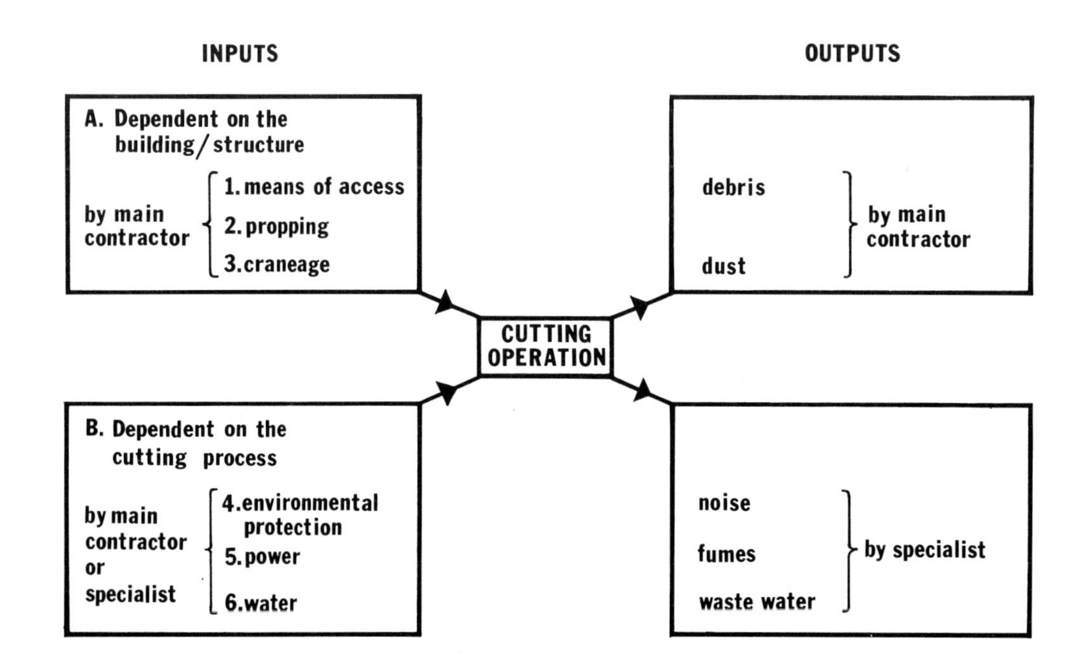

	key to requirements: definite ● occasional ▲ alternatives ○	technique																
		abrasive disc cutter	scabbling machine	floor grinder	rock drill	diamond drill	diamond saw	oxygen cutting	powder cutting torch	thermic lance	thermit powder demolition	flame spalling torch	"rocket" jet burner	water jet	concrete breaker	explosives	hydraulic burster/jack	CO2 gas expans. burster
services	water				▲	●	●					▲	●	●				
	electricity	○			○	○									○			
	drainage					●	●							●				
provisions	cylinder gases							●	●	○		●	○					
	lifting equipment						▲	▲	▲	▲	▲			▲		▲	▲	▲
	forced ventilation							▲	▲	▲								
	drainage pumps					▲	▲							▲				
	fire extinquishers							▲	●	●	●	●	●					
	system for warning site personnel										●					●		▲
	accoustic screening		▲			▲	▲						●		▲			
	screening against dust and/or fumes		●	▲					●	▲		●	▲		▲			
facilities	air compressor	○	●		○	○	○								○		○	
	liquid oxygen tanker							○	○	○		○	○					
	high pressure water pump													●				
	electric generator					○	○											
	hydraulic pump				○	○	○								○		○	

4.6 Support services, provisions and facilities

5 ENVIRONMENTAL ASPECTS AND HAZARD CONTROL

One of the most important considerations in the choice of method for cutting or breaking operations is that of the environmental effects of the work. The public interest, however that is assessed, is increasingly concerned at the impact of modern technology on people's lives. The building industry is a frequent target for criticism in this respect, and cutting and breaking operations feature prominently.

The two principal disciplinary documents in the UK are the Control of Pollution Act 1974 and the Health and Safety at Work Etc Act 1974. In principle, these deal respectively with the permissible effects of site working, and the exposure and protection of workpeople; they are particularly relevant to noise problems. These two pieces of legislation have probably dealt more thoroughly with the matter than corresponding statutes elsewhere in the world, although many countries have introduced controls to restrict noise levels.

Traditional methods of cutting and breaking presented problems of noise, vibration and fumes and/or dust, but rarely problems of water or fire. Although the techniques reviewed in this book, as currently applied, may well be less damaging overall than the traditional ones, they still may require particular measures to minimize such difficulties.

The different types of hazards, to operators and to others in or near the site and also to the structure of the building, are discussed in this chapter together with some guidance on preventive or protective measures which can sensibly be undertaken. At the end of the chapter, Table 5.2 sets out those factors which are relevant to the principal processes, with a summary of the corresponding preventive/protective measures. In order to appreciate the measures which might be necessary or possible in each case, reference should be made to the information on the particular process, in chapters 2 and 3.

5.1 HAZARDS TO OPERATORS

5.1.1 NOISE

Noise is one of the less desirable by-products of the introduction of more machinery on to construction sites. The comparative effects of the noise caused by the various types of cutting and breaking tools may be of considerable importance to the specifier and contractor. Unfortunately some of these effects are highly subjective, and those that are capable of measurement are moreover prey to a large number of variables. Table 5.1 gives general guidance on comparative values. It will be seen that only flame processes are likely to be free of the possibility of difficulty.

As has already been stated, legislation in the UK is probably more onerous in its treatment of noise factors than that in other countries. While the two principal Acts in the UK are therefore considered, the statutory requirements elsewhere, where they exist, may have similar effects. The effect of the Control of Pollution Act 1974 is not to lay down definite noise levels for a general set of conditions, but empowers local authorities to set up a pattern of working and acceptable noise levels for each particular site. In doing this, the local authority will be free to take into account any circumstances that they feel are relevant. They may impose restrictions in advance of the commencement of work, or at any time during the contract. However, the contractor or specifier may ask the local authority, in advance, to state their requirements for noise control, and they will be obliged to answer within 28 days. The notice issued by the authority may include all, or any combination of the following restrictions:

Table 5.1 Maximum noise levels of unsilenced plant in operation compared to popular cutting and breaking methods (by courtesy of BRE)

Plant type*	Sound level at 7m dB(A)	Sound power level dB(A)
Pneumatic pile driver	105–110	130–135
Scraper, 24m³	93–107	118–132
Crawler dozer, 104 kW	84–93	109–118
Crawler excavator, 68 kW	83–93	108–118
Fork-lift truck, 1·27 tonne	85–90	110–115
Diesel concrete mixer, 200 litre	83–89	108–114
Truck mixer	79–89	104–114
Tandem pedestrian vibratory roller, 6 kW	83–87	108–112
Shaft driven poker vibrator	72–87	97–112
Wheeled excavator, 45 kW	71–85	96–110

Cutting or breaking method	Sound levels at 7m dB(A)†	Sound power level dB(A)‡
Pneumatic hand-held breaker		
36kg standard	91–102	116–127
36kg silenced	83–98	108–123
36kg silenced and muffled steel	80–90	105–115
Hydraulic hand-held breaker		
30–32kg standard	82–87	107–112
Excavator mounted breaker		
Hydraulic	85–96	110–121
Thermic lancing	69	94
Diamond drill and saw	79–108	104–133
Nibbler fitted to a hydraulic excavator	76–82	101–107

* These familiar items of construction plant are included for the purpose of comparison with cutting and breaking equipment shown in the second part of the table.

† Measured with a sound level meter set for "slow" response.

‡ Sound power level is a convenient expression for describing noise-output as it does not require qualification by distance.

5.1 Fume and fire problem dramatically illustrated in this example of a Thermit powder demolition

(a) continuous noise levels which may be emitted from the site;
(b) restrictions on working hours for the site as a whole, or any particular noisy operations;
(c) peak noise levels which may not be exceeded.

The Health and Safety at Work Etc Act 1974, like the Factories Act, is a penal statute in the UK, and the penalties upon conviction of an offence are a fine, or, in some cases, imprisonment. The protection that this Act gives to workers who are required to operate in a noisy environment cannot be absolute, but taken together with the Code of Practice for reducing the exposure of employees to noise, the risk of irreversable damage to the hearing from long-term exposure to noise will be greatly reduced. Management is given a general responsibility for ensuring that the best practical methods for noise reduction are used, and that when this has been done, and the noise level remains higher than the recommended maximum of 90 dB(A), that suitable ear protectors are both available *and are worn.*

British Standard Code of Practice 5228 *Noise control on construction and demolition sites* gives a great deal of information and guidance, and lists of relevant legislation in the UK.

5.1.2 NOISE AND NUISANCE

Noise does not have to be physically injurious for it to be a nuisance. Intelligent anticipation of likely problems arising from the effects of noise upon the community have to take account of the susceptibilities of the neighbourhood. Noise may interfere with working efficiency by inducing stress, and disturbing concentration, especially where people are engaged on difficult or highly skilled work. Persistent noise levels at particularly irritating frequencies may be worse in this respect than intermittent higher noise at a different frequency. Noise may also be positive danger on site because it will make communication difficult, and may mask normally audible warning signals.

The propagation of noise through a structure or building will be almost impossible to predict. The structure itself may attenuate some frequencies and transmit others; air-conditioning trunking and pipework will transmit and effectively amplify some noise, and parts of the structure may vibrate in sympathy, generating new sounds. These are some of the effects which cause problems when buildings are being modified while they continue to be occupied.

The general view is that airborne noise is quickly attenuated especially at the higher frequencies, while noise that originates from impact or vibration of the structure itself is more difficult to contain. In this respect the noise of diamond sawing or drilling, which to the listener in the same room may be as unpleasant as that of a pneumatic breaker, may be inaudible in a distant part of the building where the breaker can still be heard.

The noise from the proper use of explosives is unlikely to be troublesome to the general public, although when combined with other effects such as vibration it may cause fear.

5.1.3 FUMES AND DUST

Fairly obviously, the principal sources of fumes are flame processes, while dust is the inescapable consequence of dry attrition. Airborne pollution of this kind can be a nuisance to operators, because it may be hazardous for them to breathe and it may also severely restrict visibility.

Thermic lancing and powder cutting of reinforced concrete will cause a considerable amount of, possibly, very dense smoke to be emitted. The smoke and fumes will not themselves be hazardous, but will probably be unpleasant for the operator. If cast-iron is being cut, the smoke may be dense enough to prevent further work anywhere but on an open site.

Although neither the dust nor the smoke which naturally arise from any of these processes is particularly harmful, this may not be true if certain kinds of finishes have been applied to the surface, or if contaminants have soaked in. It is impossible to define all the dangers, but it is essential to proceed with caution if a known health hazard has to be ground or burnt, or if there is any reason to suspect the presence of something deleterious.

A list of dusts creating health hazards is given below. The list is for guidance only and is not exhaustive. The fumes given off by many materials when heated may be more injurious than the dusts mentioned above, and in this context it is worth noting that thermic lancing and powder cutting can emit both fumes and dust.

Dusts. Health hazards result from inhalation of the following:

(i) Very dangerous
- beryllium
- silica which has been heated (calcined)
- blue asbestos

(ii) Dangerous
- other asbestos
- silica
- mixed dusts containing 20 per cent or more of free silica
- fireclay dust

(iii) Moderate risk
- mixed dust (less than 20 per cent free silica)
- talc
- mica
- kaolin
- carbides of some "hard" metals
- cotton dust and many other vegetable dusts
- graphite
- coal dust
- aluminous fireclays
- synthetic silicas (possibly low risk)

(iv) Low risk (often exposure is far too high, however)
alumina
baryteo
carborundum
cement
emery
ferrosilicon
glass (and glass fibre)
iron oxides
limestone
magnesium oxide
mineral wool
slag wool
perlite
silicates (other)
tin-ore and oxides
titanium dioxide
zinc oxide
zirconium silicate and oxide.

Fire and explosion hazards

It is well known that dust clouds can explode, given the right conditions. The following can be particularly hazardous:

(i) Metal dusts
magnesium
aluminium
cadmium
zinc
copper
iron
manganese
titanium
ferromanganese
antimony
zirconium

(ii) Other dusts
coal
dusts of vegetable origin
paper
pitch
resin
synthetic resins
fillers

Ignition sources may be any of the following:
electrical sparks
static
frictional sparks (boots, hammers, grinding wheels)
hot particles
open flames.

5.2 HAZARDS TO OTHERS ON OR NEAR THE SITE

While this is not such a serious problem as the hazards to operators themselves, or to the structure of the building, the possible danger to other workers on the site or people living nearby must be given proper consideration.

5.2.1 NOISE AND VIBRATION

The effects of noise and its nuisance have been covered in 5.1.1 and 5.1.2. They should obviously be taken into account for persons on or around the site, in addition to the operator. Some of the breaking processes also cause vibration, but fortunately the average observer should have adequate warning of the approach of damage to the building because the vibration would already be causing personal discomfort well in advance of any failure of the building.

5.2.2 FUMES AND DUST

The public and other personnel must be kept clear of all hazardous fumes or dust and the operators must be adequately protected. Non-hazardous fumes or dust may still be sufficient to cause annoyance to the public, though they ought to be dispersed quickly enough by the measures taken to protect the operators themselves (see table 5.2).

5.2.3 WATER

The presence of substantial quantities of water may also be viewed as an environmental hazard to other workers in the building as they experience the necessity of wading through puddles or avoiding the cold showers from above. However, with a little care, this particular problem can be overcome fairly easily.

5.3 HAZARDS TO THE STRUCTURE OF THE BUILDING

5.3.1 VIBRATION

Apart from the specialized cases of vibration arising from the use of some of the breaking processes or of explosives, care should be taken when using any device that has reciprocating or eccentrically (off balance) rotating mass. Such equipment will tend to set up vibration of the structure, which will be readily transmitted through the continuity of the solid material, and will re-create sound waves at points remote from the source.

Because of their mass, buildings usually vibrate at very low frequencies, of the order of 1–10 Hz, but light, relatively thin, panels can be stimulated to vibrate at audible frequencies. Damage to the building structure is caused by the combined effects of frequency and amplitude, and while this would imply the absorbtion of colossal power when considering the building as a whole, dangerous levels of vibration could easily build up in, say, a light-weight concrete floor slab from a quite modest input of energy. Heavy floor scabblers, and even heavy pneumatic breakers, used in the centre of relatively thin concrete slabs can cause vibration damage in the form of patterns of cracks, but are unlikely to bring about structural failure.

It is as well to remember that a concrete floor saw, fitted with an internal combustion engine, is also a vibrating load, and will need an additional margin for safety over the static weight. Other power sources would not have this disadvantage.

5.3.2 FUMES AND DUST

Though these do not cause structural damage, they may cause staining or other damage to building surfaces which are to remain.

5.3.3 WATER

Water may be introduced to cutting operations as a necessary coolant for diamond drilling or sawing, or in the case of water-jetting as the cutting medium itself. Without some care in the pre-planning, it is also possible to break a water pipe in an occupied building. The problem is essentially the same in these cases in principle, although not in degree. The precautions and solutions are similar, and are discussed in section 5.4 on precautions (p. 51).

5.3.4 FIRE

Generally speaking, it is only the heat/flame processes which present a fire hazard. The risk is probably greater of causing a secondary fire among combustible material which is nearby, rather than from the immediate process itself. The more powerful

5.2 The use of silenced compressors can greatly reduce the annoyance caused by plant

the process, and the higher the temperatures involved, the greater will be the risks. Therefore, powder cutting, thermic lancing, Thermit powder work, and oxy-kerosene rocket-jet shaping are all processes which merit close attention to this aspect.

5.4 PRECAUTIONS

After this review of the different types of hazard which may be encountered, the preventive and protective measures appropriate to each hazard are now discussed.

5.4.1 NOISE

The effects of noise may be mitigated in two stages; firstly by attending to the sources of noise, and reducing them as far as is practicable, and secondly by then containing within the immediate area whatever noise is emitted. Within stage one, some very simple improvements to any of the machines can be made by ensuring that there are no loose nuts and bolts, or badly worn components. Then consideration should be given to the use of muffled tool steels on concrete breakers, the use of sound deadened diamond blades on saws (which reduce the "ringing" noise from the larger blades as they cut), and the fitting of silencers or mufflers to the exhausts of all pneumatic motors.

Once everything has reasonably been done to cut down on the generation of noise, then the second stage, the control of its propagation, may be tackled. This will involve building enclosures around the noisy equipment, ideally using material approved for its attenuating properties, or if special material is not available or is too costly, then the thickest possible plywood or blockboard screens. Precise guidance cannot be given in this respect, and experience, or trial and error, are often necessary.

Table 5.2 Process hazard control

Process	Noise	Vibration	Fumes/dust	Water	Fire
"Soft" cutting tools	○	○	◑		
"Hard" cutting tools (diamond drill or saw)	◑			◑	
Heat/flame processes:					
Oxy-acetylene			○		○
Thermic lance			●		●
Powder cutting			◑		●
Thermit powder burning			◑		●
Oxy-kerosene "rocket jet" shaping	●		●		●
Flame spalling			◑		◑
Water-jet cutting	○			●	
Impact/percussion breaking	●	●	●		
Explosives	◑*	◑*	◑		

Sensible preventive measures					
● major factor	no practicable reduction method; operators must wear ear protectors	no sensible reduction or protection methods available	heavy screens, careful sealing, forced ventilation, breathing aids or filtration	elaborate screening, collection and pumping	clear area of all combustibles; protect heat sensitive materials; provide appropirate extinguishers
◑ moderate nuisance	silencing jackets, muffled tools, and site plant remotely; ear protection	,,	light screens, local fume extraction	simple protection, pumping or drainage or	as above, but more localized
○ little problem	site plant away from sensitive areas	,,	ensure good natural ventilation	local collection, possibly re-cycling	simple precautions water or sand available

*Of short duration

5.4.2 VIBRATION

It will be seen from table 5.2 that in general terms diamond tools, heat/flame processes and water-jet cutting are reasonably free from vibration problems. Those processes which do cause such problems, to whatever degree, are not capable of sensible treatment to deal with it, and the specifier or contractor must select the process accordingly.

5.4.3 FUMES AND DUST

Confined situations can be greatly improved by using a ducted extraction system with a specially designed inlet arrangement to allow the smoke to be drawn off as quickly as possible, before it has a chance to disperse; the discharge of such a system would be led outside the building. However, even at this stage, it may still be sufficiently noxious to infringe statutory regulations, or cause a nuisance. It is generally better to arrange for the site to be well ventilated by a large volume of induced air, thus causing the smoke to be dispersed before it passes into the surroundings.
Cement dust from concrete cutting or breaking is a common occurence, and although it is not a health risk, over-exposure should be avoided. The operators of scabbling machines and dry grinders should be wearing face masks incorporating dust filters. Where this work has to be done in buildings that are in use the contractor should be instructed to make the work area dust proof with screens of plastic sheeting securely held in position, perhaps by a framework of timber battens.
5.1.3 gives guidance on those materials which are particularly hazardous, either as dust or in fumes, either as health risk from inhalation, or as fire and explosion hazards.

5.4.4 WATER

The working area must be closely examined, and the nature of the structure determined before the water control requirements can be decided upon. Pot floors and hollow core beams and slabs are especially difficult to deal with, for as soon as the cavity is broken into water will travel throughout any continuous void system. Water control is usually achieved by stapling polythene sheeting to light battens fixed to the workface, which prevents that from becoming soaked, and also directs the water into a suitable collection area. Some spray may be thrown around by a revolving blade or bit, and this may have to be collected by suitably positioned sheeting. The collection point may simply be a sheet metal tray, or alternatively if the work is extensive, a single course of bricks may be laid in mortar on the floor to form the walls of a temporary pond, the interior of which is painted with a bituminous coating. The water is usually pumped from this point with pneumatic or electric pumps, to waste.
A more elaborate apparatus is shown in 5.4, in which a vacuum pick-up is used to collect the water close to its egress from the blade or bit. This system also allows the cutting debris to be separated from the water, and the clean water to be re-used for cooling, or run to waste without the risk of blocking any drains with solid material. Vacuum water recovery may also be used to overcome a secondary nuisance arising from the free drainage of

5.3 Metal tray and electric pump for collecting waste cooling water

debris-laden water, which might leave a permanent stain on the workface as the slurry dries out.

5.4.5 FIRE

Some simple precautions should be taken in all cases. All removable combustible materials should be cleared from the work area, and all heat-sensitive materials must be protected with asbestos blankets, metal screens, and the like. A supply of water or sand should be available near the work area for the less powerful processes, but those itemized earlier as producing intense heat must be provided with fire extinguishers; if unusual materials or circumstances exist, then specialist advice is necessary to ensure that the extinguishers are appropriate.

In cases where hazardous or combustible materials cannot be removed from the area, and where protective measures are impractical, alternative cutting processes should be selected. (See also the list on pp. 49–50 for guidance on substances which may present particular fire or explosion hazards when in dust form.)

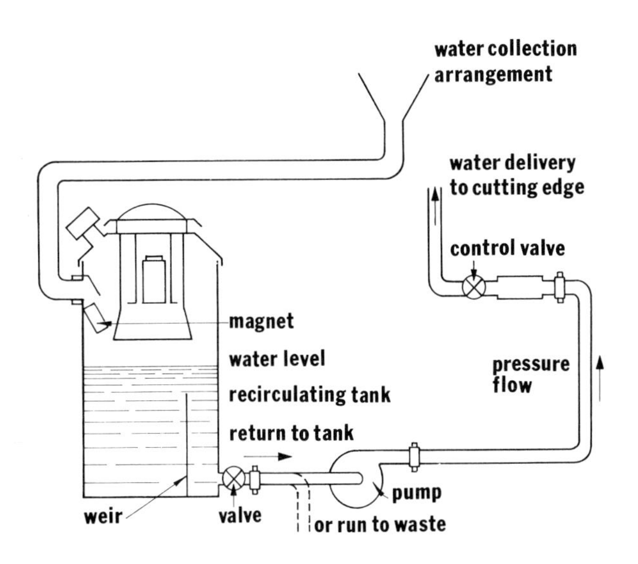

5.4 A more sophisticated system for water recovery and/or recirculation. Cooling water is fed to the diamond tool and is then drawn back through a water collection unit to a recirculating tank where metal filings and abrasive particles are separated; clear water may then be fed back to the cutting face or run to waste

6 COSTS AND OBJECTIVES

Chapters 2 and 3 have reviewed the various techniques and processes commonly available, with their respective applications. However, it will be clear from chapters 4 and 5 that a number of other factors must be considered, in addition to the simple technical ability of a process to achieve the desired result.

The influences of so many possible constraints and requirements on the economic performance in a particular case are obviously complex, and are certainly too variable to be capable of simple summary. The experience of specialist contractors is probably the best guide available to the comparative overall economy of the various alternatives; nonetheless it will be useful to consider the way that the cost of the work is predicted, with the corresponding build-up to the quoted rates for the different processes.

It will also have become clear that there is a wide range of objectives to which the processes may variously be applied. The common ones will be fairly apparent, but it is important to recognize the adaptability of much of the equipment and techniques; there are few objectives which should be discounted without careful study of the possibilities of applying a process, or combination of processes, to the particular situation.

Tables 6.4 and 6.5 set out the common methods of achieving various objectives, and techniques which are appropriate to the particular cutting jobs; broad cost comparisons are given. It must be appreciated that this cannot be an exhaustive or precise summary, as the economy of performance in one situation can be reversed in a different situation, and there are many instances when acceptable results can be obtained by the use of the simpler methods; therefore it is impossible to give anything but very general guidance.

6.1 COST CONTROL OF CUTTING AND BREAKING WORKS

In a sense, the designer or engineer who specifies a particular procedure is also the buyer, and as such, he needs to know not only what suppliers are charging but also what the real costs of the work should be. If the buyer has an appreciation of the various cost elements which go into the build-up of the price, he will be enabled to:

- make more objective selection from competing quotations;
- monitor the work in progress, and take early action to regulate costs which are going astray;
- deal more confidently with prices for variation orders and extras.

It is important to remember that the prices quoted in the UK by specialist contractors in the field of in-situ cutting will probably be subject to a much wider spread than those quoted by other sub-contractors in well-established trades. This may not be true in those parts of the world where particular techniques have become established and now operate in a more disciplined market, for instance the West Coast of the USA, where diamond sawing and drilling are well understood, and the market is extremely competitive. However, taking the UK market, and the provision of less well-known techniques as examples, this wide spread of prices for the same job might be anticipated because of the number of subjective pricing judgments that the specialist will try to include. His price will reflect such factors as:

- the anticipated direct cost of the work;
- how tightly the work is likely to be controlled, and the scope for escalating the contract costs;
- the degree of uncertainty affecting the execution of the contract, and an assessment of the risks associated with any unknown factors;
- whether there is any competition, and how they are likely to price;
- how important it is for him to win the contract at a particular time.

It may be seen from these how important it is for the specifier to have an understanding of the pricing mechanism, if the work is to be planned and controlled. This should not be confused with the objective of obtaining too low prices for work, because this would eventually curtail the availability of truly competitive specialist cutting services. The techniques will only be developed, and their applications extended, against a background of stable, sensible prices, which give the contractor an adequate return, and at the same time instil confidence in the user. Silly, low prices cause contractors to withdraw in the middle of unprofitable jobs, or to disrupt progress in other ways, possibly by switching resources to other less "loss-making" sites. In the long run, no one could benefit from this situation.

The first, and most important, factor noted for the price build-up is the direct cost of carrying out the task; this cost will be analysed shortly. The second is the degree of control to be exercised by the specifier, and this book should enable this to be better executed. The remaining factors are purely commercial judgments, and are not within the scope of this study; their influence, however, is obvious.

In order to establish the various known direct costs included in the price build-up, these costs are broken down to their constituent factors:

1 Time
2 Materials
3 Down-time
4 Logistics

6.1.1 TIME

Time is convertible directly into money terms through the application of labour rates, and plant and equipment use rates. Plant and equipment rates are related not only to the capital cost of the equipment, but also to the utilization that the con-

6.1 Four stages in the demolition of a steel and concrete coal handling plant using Thermit powder to burn through the supports. After the preparatory work, the cutting process is very fast

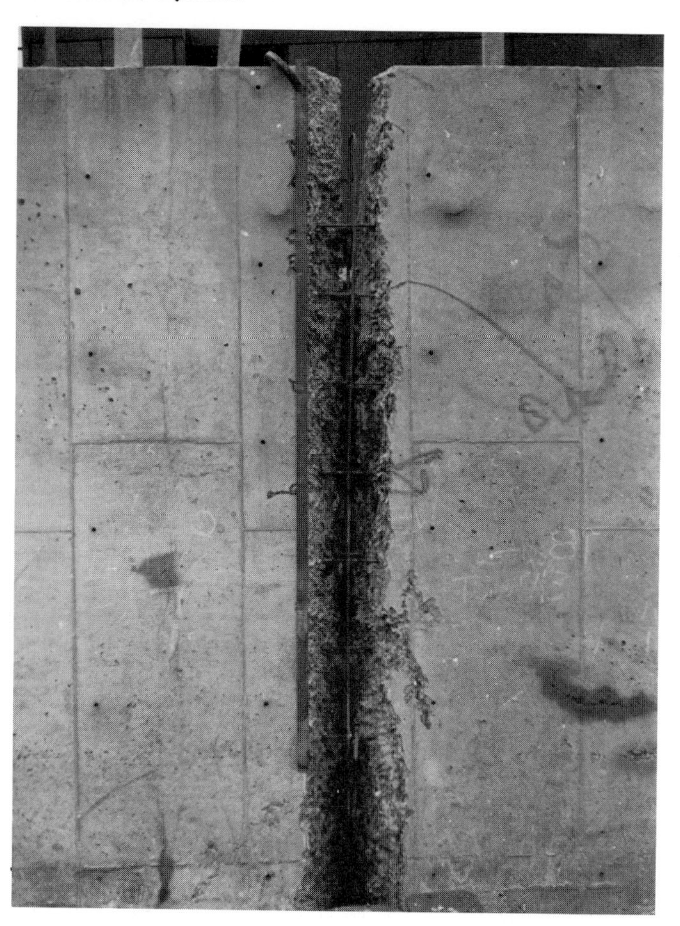

6.2 The 3·5m × 300mm cut formed in 7½ hours, by water jetting

tractor can expect to obtain; in other words, the more specialized the item, the less frequently it may be required, and the higher the cost that has to be recovered during those periods when it is employed. This incidentally illustrates the problem of comparing alternative methods simply on the basis of apparent costs, because, in general, the use of higher cost plant results in reduced working time, but the point at which the balance is struck between reduced working time, and increased hourly recovery rates may be difficult to identify. Many specialist contractors issue schedules of rates for all their major items of equipment, and daily rates for hiring operators. Down-time (discussed below) should be considered in conjunction with time.

6.1.2 MATERIALS

The second of the two cost factors contained in the work element is material. This may be of major significance in the overall cost because many of the cutting procedures involve the consumption of high cost items, such as diamond blades or drill bits, or large *volumes* of materials, such as thermic lances, powder mixtures or gases.

The mechanical cutting methods consume material through wear on the cutting edges; as it is likely that a diamond blade or drill bit may have a useful life spanning several separate small contracts, it is fair to assume that only the wear attributable directly to the performance of one contract will be charged to that contract. The relationship between the wear on a diamond cutting edge and the parameters of the cutting situation are extremely complex (see pp. 59–60).

Forecasting the consumption of abrasive discs or tungsten carbide tools may not seem at first sight to be as important, as these are comparatively low cost items. However, they may wear out very quickly in some circumstances, and their total cost may then outstrip that of an initially more expensive cutting tool. Manufacturers' literature usually contains information on the expected life of the products under a variety of conditions; however, this may sometimes have been obtained by laboratory tests and similar results may not be obtainable under typical site conditions. (The short life of some of the low cost cutting materials can generate further costs by increasing the down-time needed to change blades at frequent intervals.)

In contrast, the consumable costs of the flame processes, for example thermic lancing, powder cutting, and flame spalling relate to the total consumption of the various sources of energy. Thermic lances burn rapidly, and although their unit cost is low, the aggregated cost in comparison with the amount of work done is high (ie burning rate around 2m/min (6–7ft), penetration around 200mm/min (8in)). Fortunately, thermic lancing is a reasonably predictable process, and simple relationships between lance and oxygen consumption and work done are acceptable for commercial purposes.

Such assumptions cannot be made about the other flame processes, notably powder cutting and flame spalling. Variations in the types of concrete can have considerable effect upon the performance, and very often nothing short of a trial pass can establish the data for estimating the cost of the job. Once again, manufacturers' literature can be used to find the consumption of gases, or powder, in relation to running time, but not the material being worked.

Finally, an often overlooked consumable cost is that of providing power. This is almost invariably in the form of fuel cost for compressors, hydraulic pumps, or electricity generators, and must be gauged from the power rating of the prime-mover.

6.1.3 DOWN-TIME

The non-productive time or down-time should be known, or a reasonable forecast be made. Down-time can comprise a multitude of activities which will differ widely from situation to situation, but which may be taken to include such things as time for setting up the machinery on site, changing the position of equipment, changing blades or bits, breaking out detached material, and clearing site.

For estimating fairly standard procedures, particularly on jobs of sufficient duration to allow a pattern of working to become established, the usual practice is to add a fixed percentage of extra time on to the cutting time, itself calculated from the performance characteristics of the process, in order to cover the expected down-time. This percentage is likely to lie between an additional 50 and 100 per cent, depending upon the methods to be used and the complexity of the job.

6.1.4 LOGISTICS

The logistical element of the total cost is the sum of all the costs that are likely to be incurred by the specialist contractor in placing his operators and their equipment in a position where they are able to perform the work, and subsequently recovering in readiness for the next contract. The logistical costs are of course almost infinitely variable, but as they will be made up from readily established cost factors, there should never be too much difficulty in anticipating the rough order of cost of working on a particular site. Some of the following elements could be included in this assessment: transportation of equipment, travelling time and costs for operators, subsistence and accommodation, site storage, and communications.

6.1.5 OVERALL COST

It would be unrealistic to treat this information as anything more precise than a general guide to the method of price build-up for a particular job. There will always be procedures and situations

6.3 Stitch drilling mass concrete to form an opening in a water intake shaft near Chicago

6.4 The completed opening showing the scalloped edge to be made good

6.5 Thermic lancing a hole through a redundant rc beam

which will make such an analysis meaningless and where the right price for the job must simply be the price that the client is willing to pay for the service; in such a case time and materials can be irrelevancies. For *most* work the pricing mechanism outlined will apply. Without a doubt competitive pricing within an informed market will provide the best basis for comparison.

Cost comparisons between the various methods reviewed and traditional ones are extremely difficult to make because of the enormous number of variables which may have to be taken into account. Several researchers have attempted studies to compare the costs of alternative routes to the same objective but little confidence can be placed in the results. One of the most detailed studies investigated the cost of forming round and rectangular holes in concrete, and set out to compare all the alternative forming or "grout-box" methods for forming the hole, as against all the alternative "breakthrough" cutting procedures. The result showed a substantial advantage in the use of diamond drilling, but too many factors had been ignored to put the findings beyond some dispute. Tables 6.1–6.3 illustrate the costs which were established in the study at 1972 price levels.

As an illustration of the cost relationships which exist in the *diamond* processes two graphs **6.6** and **6.7** are included. Graph **6.6** illustrates the relationship between the diameter of hole being drilled and the cost per metre (measured in an arbitrary unit); this assumes a ratio of labour cost to drilling bit consumption, which may vary a little in practice. It is particularly interesting to note that there is an optimum size of hole, above and below which size the costs can increase significantly. When stitch-drilling a line of holes, for example, where the individual hole size is usually unimportant, considerable cost savings may be possible by the selection of a size which lies within the minimum cost band. The area between the two curves is indicative of the spread of performance costs which are to be expected from variations in the sand and aggregate qualities, and in the amount of steel.

Graph **6.7** illustrates the relationship between the area of cut by diamond sawing, ie the length of cut multiplied by the depth, and the cost measured in a similar arbitrary unit to that used for the drilling costs.

The transition from unreinforced materials to steel reinforced materials cannot be shown as a continuous curve because the performance changes abruptly at this point and there is some evidence from measurements taken in the field to indicate that a different relationship exists.

Although both these graphs are simplified pictures of the situations which can actually exist, graph **6.7** is open to further error in that constant values for the area of cut can contain a number of variations in the ratio of depth of cut to length. These

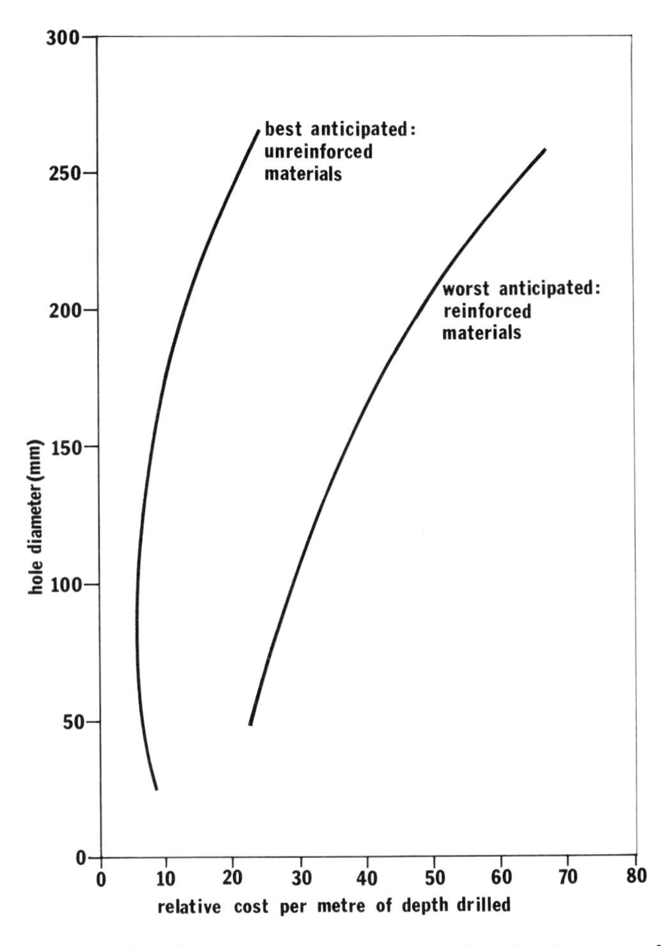

6.6 Diamond drilling costs in concrete related to the diameter of the hole

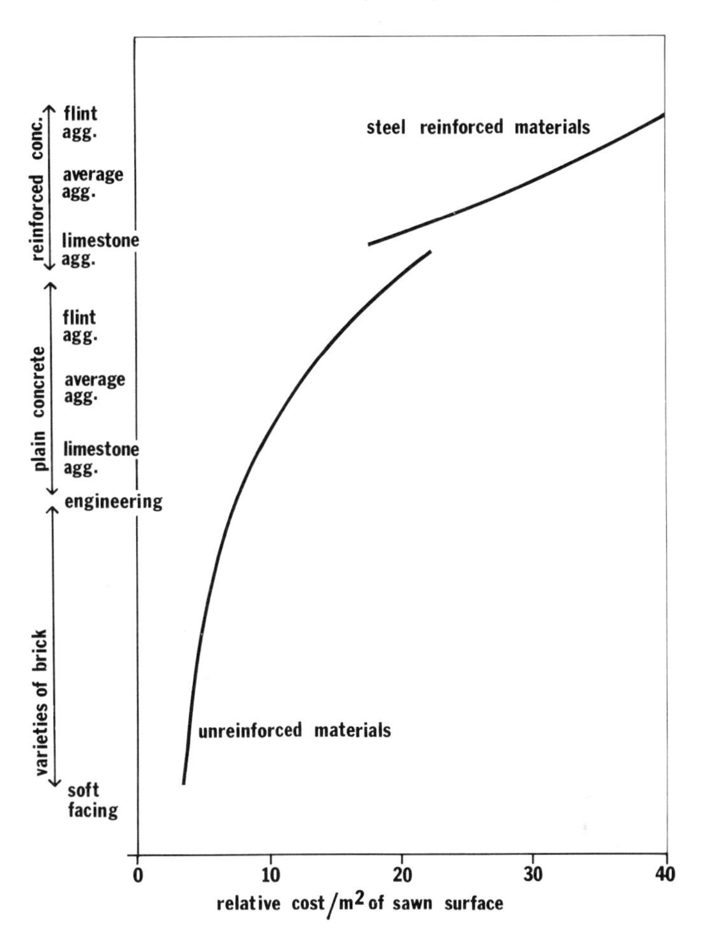

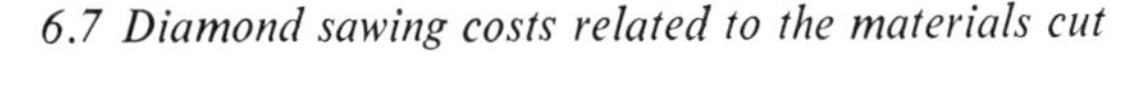
6.7 Diamond sawing costs related to the materials cut

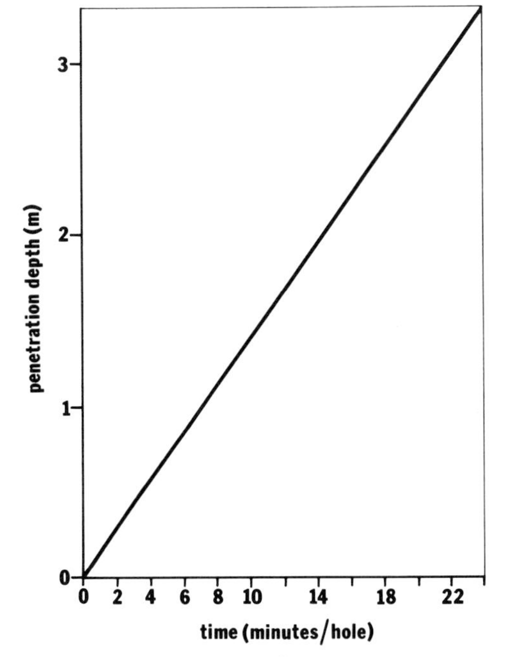

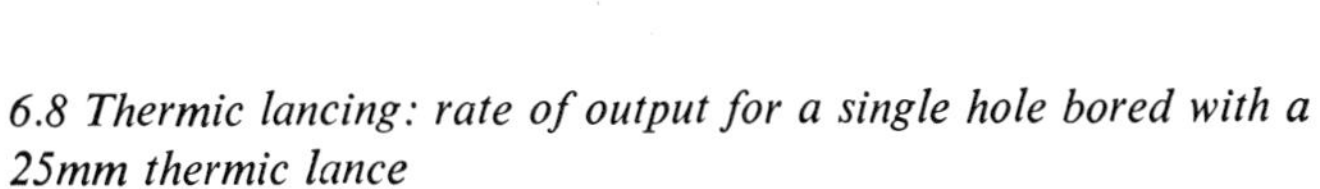
6.8 Thermic lancing: rate of output for a single hole bored with a 25mm thermic lance

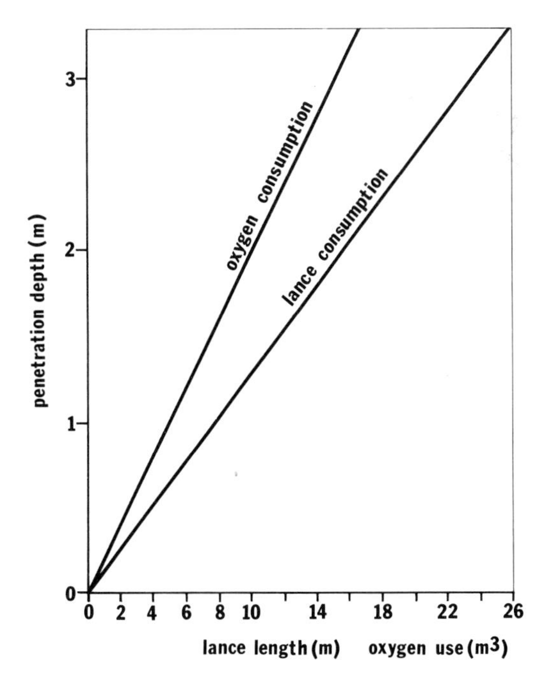

6.9 Thermic lancing: lance and oxygen use for a single hole bored with a 25mm thermic lance

variations can affect the total cost in several ways, by increasing the number of passes necessary to achieve the depth with the consequent increase in the non-productive time, and by greatly altering the efficiency of the sawing machine for example.

With regard to other common processes, there are too many variables affecting performance to be able to draw meaningful, but simple, performance comparisons. For example, it is impossible to compare the cost of thermic lancing on the same basis as that used for the diamond processes because, as has already been explained, the mechanism by which cutting is achieved is very

Table 6.1 Influence of size of hole on drilling costs using diamond core-drilling (see Note opposite)

Costs involved in this method of making holes were obtained from invoices, over a recent one-half year period, of a large drilling contractor. Listed below in tabulated form are total costs in the most popular sizes.

Diameter mm	in	Thickness cm	in	Size reinforcement mm	in	Total price £	US $
Vertical drilling							
52	2	10	4	16	5/8	3·93	7·08
52	2	18	7	16	5/8	6·76	12·15
82	3 1/4	10	4	16	5/8	4·62	8·32
82	3 1/4	18	7	16	5/8	6·93	12·45
100	4	10	4	16	5/8	4·00	7·20
100	4	18	7	16	5/8	7·10	12·78
122	4 3/4	10	4	6	1/4	4·43	8·00
122	4 3/4	18	7	16	5/8	8·64	15·55
162	6 3/8	10	4	6	1/4	5·55	10·00
162	6 3/8	18	7	16	5/8	9·24	16·60
Horizontal drilling including 3.50m (10ft) high scaffolding							
52	2	20	7 7/8	16	5/8	8·90	16·00
82	3 1/4	20	7 7/8	16	5/8	9·24	16·60
162	6 3/8	20	7 7/8	16	5/8	14·15	25·50

Table 6.2 Typical build-up of costs, for hole drilling using alternative methods

Trial drilling		**Percussion drill**	**Thermic lance**	
Thickness of concrete	cm	16	16	
Flooring screeding	cm	4	4	
Total	cm	20	20	
Oxygen consumption	litre/min		500	
Diameter of hole	cm	4·0	4·2–4·5	
Drilling time	min	approx 40·0	approx 2·0	
Output	cm/min	0·5	10·0	
Lance consumption	m/hole		1·2	
Material	£			1·05
Persons		2	3	
Wages	£3·31/hour	2 × 2·21	3 × 11p	33p
Cost per hole	£	4·42		1·38
		Probable total costs (per similar task of 760 holes)		
Trial drilling		**Percussion drill**	**Thermic lance**	
Hours work der day		10	10	
Holes per day		15	150	
Persons		2	3	
Workdays		50·76	5·076	
Hours	£3·31/hour	3369	505	
Materials		929	921	
Total cost		£4298 ($7750)	£1426 ($2575)	

Table 6.3 Comparison of cost of pre-forming holes in concrete or subsequently cutting out

	By formwork or grout-box						**By breakthrough cutting**				
	Formwork to be removed			**Permanent formwork**			**Heat/flame processes**		**Mechanical processes**		
	Timber	Sheet metal	Expanded polystyrene (styrofoam)	Sheet metal	Plastics	Asbestos-cement	Thermic lance (burning bar)	Powder/oxygen lance	Pneumatic concrete breaker	Hammer drills	Diamond core-drilling
	Shape of hole produced, and how frequently										
Common	□	□	□	○	○	□	○	○	○	○	○
Rare		○		□		○			□		□
Cost	Similar, Average £9·50 ($17·00)								£14·40 $26·00		£8·30 $15·00

* Holes by pneumatic concrete breaker include £4·50 for making good oversize opening.

6.10 1·5m thick concrete step-cut by diamond saw

6.11 This permanent concrete casting mould is being modified by diamond sawing to increase its width and to part-chamfer the entry. At this stage the extreme delicacy and accuracy of the cutting can be appreciated

different. For costing purposes, a simple linear relationship between the depth of hole bored by a single lance and the labour costs, and also depth and the consumption of lance and oxygen can be used. The relationship can be assumed to hold true for use on a wide variety of materials because lancing is not sensitive in this respect. However, lancing does not produce dimensionally accurate holes, nor is the mechanism for achieving a cut a precise one, and therefore the ancillary costs previously discussed are of particular importance. The two graphs **6.8** and **6.9** illustrate the simple relationships outlined.

Note: Examples of cost comparisons made on alternative methods for simple holes (see opposite)

The three tables opposite are based upon the researches of Professor Klaus Simon, of the Technical University of Braunschweig, and others, and were published in 1972.

The values quoted relate to the German market *in that year* and obviously cannot serve as a guide to current prices in other markets. The tables do, however, illustrate the complexity of analysis needed to make a true comparison, in addition to their interest for the comparisons that they contain.

6.2 OBJECTIVES

Any building project, whether new construction or alteration, will almost certainly have elements which will require, or would benefit from, the application of the techniques which have been reviewed. The designer/specifier should not only consider the more familiar building operations which modern techniques aid, but should also expand his horizons of feasibility by applying the techniques more imaginatively, or on a larger scale. The case studies in chapter 8 illustrate a number of ways in which the more common ones have been used, and table 6.4 shows the methods of achieving known objectives. Several other less known objectives are discussed at the end of this chapter.

In considering the work for which the cutting and breaking techniques will be useful or essential, it is helpful to bear in mind that the nature of the work, as defined by the scope of the operation in relation to the material, can range over:

1 Removal of members;
2 Breakthrough, ie forming perforations of any size;
3 Partial penetration, eg forming pockets, grooves, joints etc;
4 Surface treatments.

The common objectives covered in table 6.4 can be viewed in these categories.

Having established, in principle, the scope of the work envisaged, the objective must be considered in relation to the total structure, and the overall contract work. For example, the removal of the wing of a building could be thought of as total demolition, but so far as the *whole structure* is concerned, it must be treated as controlled, *local* demolition. This ultimate objective can then be converted into a number of cutting and breaking tasks.

At this stage some definition can be introduced by identifying the materials to be cut, the dimensions of the work, and the accuracy required.

Attention should now be focussed on the obvious constraints surrounding the performance of the tasks. It is impractical to attempt to give an exhaustive list of the constraints that could be encountered, but the following aspects are worth mentioning.

- site conditions and environment
- location of workface
- presence and activities of personnel
- accuracy required
- timing of work.

This exercise will usually suffice to eliminate some procedures and to indicate that others seem to be potential solutions. If a major project is being planned, it may be advantageous then to call for budget prices for the acceptable alternative methods. These would provide a basis for further comparisons which would allow estimated values to be applied to the marginal benefits of each particular method in the specific circumstances of that project. For example, it might be practical to think of lifting out complete pieces of redundant structure for re-use elsewhere; while not being directly necessary, there might be a benefit in using a method such as diamond sawing, which gives clean detachment from the remaining mass.

Although the use of over-sophisticated methods (at greatly increased cost) should never be proposed on the grounds that they may provide additional benefits, or margins of safety, above the level that professional advice deems necessary, there will be instances where their use may make substantial savings possible in apparently remote parts of the project.

In addition to the more commonly encountered objectives, it may be of interest to consider some recent applications, which have been made economically feasible by sophisticated techniques.

Table 6.4 Methods of achieving some of the common objectives

General objective		Method	Material to be cut	Reference in table 6.5
Forming wall openings (eg for windows and doors)	(a) crude cutting (requires subsequent dressing)	concrete breaker	rc/brick	2/3 (a) (i)
		drill and burst	rc/brick	2/3 (b) (ii)
		stitch drilling	rc/very thick brick	2/3 (b) (i)
		thermic lance	rc/brick	2/3 (b) (iii)
	(b) precision cutting	diamond sawing	rc/brick	2/3 (c) (i)
		close stitch drilling	rc	2 (c) (ii)
		abrasive disc	thin brick only	3 (c) (iii)
		powder cutting	rc/brick	2 (c) (iii)
Forming floor openings (eg lift shafts and staircases)	(a) crude cutting (requires subsequent dressing)	concrete breaker	rc	2 (a) (i)
		stitch drilling	rc	2 (b) (i)
		thermic lance	rc	2 (b) (iii)
	(b) precision cutting	diamond sawing	rc	2 (c) (i)
		close stitch drilling	rc	2 (c) (ii)
		powder cutting	rc	2 (c) (iii)
Forming service holes	(a) larger holes or holes in non-sensitive areas	rock drill	rc/brick	2/3 (d) (i)
		diamond drill	rc/brick	2/3 (d) (ii)
		thermic lance	rc/brick	2/3 (d) (iii)
	(b) small circular holes in sensitive areas	diamond drill	rc/brick	2/3 (d) (ii)
Removal of redundant obstructive members		concrete breaker	rc	2 (a) (i)
		thermic lance		2 (b) (iii)
		powder cutting		2 (b) (v)
		diamond sawing		2 (c) (i)
Exposing reinforcement for extension of rc members		concrete breaker	rc	2 (a) (i)
		water jetting	rc	2 (b) (iv)
Holes for fixing (eg cladding panels and secondary fixings)		rock drill	rc/brick	2/3 (d) (i)
		diamond drill		2/3 (d) (ii)
		thermic lance		2/3 (d) (iii)
Holes for structural attachment (eg strengthening floors and adding members for additional floors)		rock drill	rc	2 (d) (i)
		diamond drill		2 (d) (ii)
Forming joints (eg sawn joints in concrete floors and joints in parapets)		diamond sawing	rc/brick	2/3 (c) (i)
		abrasive disc	brick	3 (c) (iii)
Grooves, chases etc (for cables, weatherings, dummy joints)		diamond sawing	rc/brick	2/3 (c) (i)
		abrasive disc	brick	3 (c) (iii)
		tungsten carbide tools	rc/brick	
Surface treatment	(a) cleaning	water jet	stone	1 (c) (v)
		flame cleaning	concrete	1 (c) (vi)
	(b) aggregate exposure and roughening	water jet	concrete	1 (c) (v)
		flame spalling		1 (c) (vi)
		tungsten carbide		1 (c) (ii) and (iii)
	(c) levelling	grinding	concrete	1 (c) (i)
		diamond bump cutter		1 (c) (iv)
	(d) extensive reduction and roughening	mechanical scabbling	concrete	1 (c) (iii)

6.2.1 CUTTING HOLES IN NEW CONSTRUCTION

The facility to cut holes in a building in order to insert services or fixings (or even larger ones for stairs, etc) is well known. However, the potential savings in time and money from forming all such holes after construction of the structure are less widely appreciated.

The savings in time arise from the ability to start construction before the precise layout of the mechanical and electrical services or fixtures and fittings is known; there is a further benefit in that variations and adjustments can be made right up to the time of cutting the holes. The structural engineering consultant/designer will be very pleased to be relieved of the necessity of showing all the holes on the structural drawings, and frequently having to change them.

Savings in money are obtained from the potentially reduced direct cost of the work if it is properly specified and controlled and from the avoidance of mistakes at the stage of building the structure.

An increasing number of buildings in the UK and in Europe are being planned on this basis; in the USA the procedure is quite widely used, and the structure of a recent tower office building in Chicago has been erected with no such holes or pockets but

6.12a,b The first stages of enlarging a door opening by diamond stitch drilling

with the expectation of cutting over 11,000 of them at a later stage.

It obviously is necessary for the structure to be designed with sufficient tolerance to enable the holes to be cut with reasonable flexibility, even though some inhibitions may well be necessary. Clearly there may be considerable structural difficulty in permitting any degree of flexibility in positioning very large openings, for staircases perhaps, and the benefits in such cases may be limited to very special circumstances.

6.2.2 URBAN IMPROVEMENT

Tree planting

The city of San Francisco is a major tourist attraction which has been the subject of a great deal of imaginative refurbishment. Recently, the appearance of many of the streets has been improved by planting trees in rectangular beds along the sidewalks. Most of the sidewalks in the city are constructed of strip laid concrete, in contrast to practice in the UK where precast slabs are commonplace One metre square apertures were cut in the concrete, using diamond blades in small floor saws. These neatly finished openings did not require edging, and the remaining concrete was undisturbed.

Wheelchair ramps

As a part of the general movement to improve facilities for the disabled, many cities in the USA are installing wheelchair ramps at kerbsides. City authorities have found that these are easily and speedily cut out using diamond bladed saws. The concrete of the kerbsides and footways may be friable, or not of particularly good quality, and it is probable that if percussion tools were used to form the openings there would be random cracking spreading into the surrounding concrete.

Drainage gullies

Where footways are extended, or road surfaces renewed, it may be economic to saw-cut gulley points in kerbs or surfaces. This technique has been extensively used in Oxford Street in London where wide pedestrian areas have been formed to incorporate the original footways, and the new drainage channels discharge into the original gullies, necessitating cutting the existing kerbs.

6.2.3 DECORATIVE FINISHES

An immense variety of techniques have been developed to improve, and relieve the normally drab appearance of shuttered or trowelled concrete. The reader will be familiar with the established techniques, but may be less aware of the possibilities of using some of the methods and equipment that have been discussed to this end.

Diamond sawing, for instance, has been used with great effect on pavement areas to produce patterns of straight, intersecting grooves which are not only decorative but also functional; the patterns can be arranged to intersect at drainage gulleys, and therefore help to dispose of surface water.

Similar treatment has been applied to concrete walls, both of

Table 6.5 Operations possible with the tools available

Objective	Technique	Cost comparison	Advantages	Disadvantages
1 Material: MASS CONCRETE				
(a) Local destruction or breaking out	(i) concrete breaker	cheap	reduces debris to small pieces, aids removal	very slow; noise, dust and vibrations
	(ii) explosives	cheap	fast, few restrictions except for licence. Simple equipment, remotely initiated	site evacuation; shock effects. Considerable experience required to get satisfactory results
	(iii) demolition ball	cheap	fast and effective on open sites	destruction may be difficult to localise
	(iv) tractor percussion hydraulic breaker	moderate	good output	noise, dust and vibration; requires considerable space to manoeuvre
	(v) carbon dioxide gas expansion bursters	moderate but depends on cost of making placement holes	fast, few restrictions on use, and can be used in confined spaces	produces major cracking but leaves pieces difficult to dislodge
	(vi) hydraulic bursters and jacks	moderate, dependent on placement holes	effective in confined space, no shock, quiet	fairly slow, and often needs further processes to dislodge debris
(b) Holes, for fixings or placement of charges etc for breaking	(i) rock drill	cheap if manually operated	fast	difficult to maintain high production rates over large number of holes without mechanically aided feed
	(ii) thermic lance	moderate	fast	smoke, fumes, fire hazard; less efficient vertically
(c) Surface treatment (not necessarily dependent on mass)	(i) grinding	cheap to moderate	good finish	very limited material removal, not suited for rectifying major uneveness
	(ii) tungsten carbide flail	moderate	produces non-skid finishes	no levelling. Unsuitable for suspended floors
	(iii) mechanical scabbling	cheap	simple; considerable reduction possible	vibration, noise and dust. Unsuitable for light, or suspended floors
	(iv) diamond blade machines	moderate	machines available to produce accurate running surfaces and skid resistant patterns	restricted to large traffic areas
	(v) water jet	moderate	much variation in amount of material removed	water and debris
	(vi) flame spalling torch	expensive	flexible, can be used in any plane. Leaves inert surface, in variety of colours/textures	heat and fumes
2 Material: REINFORCED CONCRETE				
(a) Local destruction or breaking out	(i) as for mass concrete supplemented by a method for cutting steel			
	(ii) hydraulic jacks in lanced holes	moderate	fast, quiet and little shock	their size limits application as does the considerable relative movement needed before reinforcement breaks unless steel is landed out separately
(b) Crude cutting	(i) stitch drilling (interlocking diamond drilling, widely spaced)	moderate to expensive	quiet, no vibration or dust. Very accurate. Will cut the reinforcement. The only method for great thicknesses	very slow; leaves ragged edge to be made good. Requires water and drainage
	(ii) bursters in rock and diamond drilled holes	moderate	fairly fast if conditions suitable, especially if reinforcement is minimal	unpredictable; often difficult to dislodge debris. Steel may need to be cut separately
	(iii) thermic lance	moderate	fast. Cuts any reinforcement with the concrete (up to 6m). No noise, dust or vibration	smoke and fire hazard
	(iv) water jet	moderate to expensive	completely reduces concrete to slurry plus aggregate; leaves steel undamaged but exposed	very slow, wet; hazard from flying debris. Specialized application
	(v) powder cutting	moderate to expensive	can be used in restricted space instead of lance, good surface finish	very slow; difficult to use in damp condition

Table 6.5 (continued) Operations possible with the tools available

Objective	Technique	Cost comparison	Advantages	Disadvantages
(c) Precision cutting	(i) diamond sawing	moderate to expensive	no dust or vibration, fairly quiet; accurate and needs little making good. Cuts steel with concrete	cut limited in depth to about 400mm with circular saw blades unless step-cut and broken out. Requires water and drainage
	(ii) stitch drilling (interlocking diamond drilled holes, closely spaced)	expensive	quiet; no dust or vibration. Will penetrate over 4m with accuracy	slow. Requires water and drainage. Leaves slightly "scalloped" edge to be made good
	(iii) powder cutting	expensive	quiet; no vibration. Dry	slow; finish inferior to diamond sawing
(d) Holes for fixings, services etc, placements for chargers, bursters etc	(i) rock drill	cheap	fast if steel can be avoided. Holes are good for most mechanical fixings	noise, dust. Will not drill steel. Through holes will spall at exit; possibility of other damage
	(ii) diamond drill	expensive	no dust or vibration; quiet. Neat accurate holes	slow. Requires water and drainage. Holes may need roughening for anchorages
	(iii) Thermic lance	moderate	no dust or vibration; quiet, fast	hole is tapered and irregular. Surrounding material may be degraded. Smoke and fire hazard
(e) Core drilling	(i) diamond drill	no alternative	equipment available to extract undamaged cores	
3 Material: BRICKWORK (all grades)				
(a) Local destruction or breaking out	(i) as for mass concrete			
(b) Crude cutting	(i) generally as for concrete	all processes cheaper than in rc	value of vibration-less processes in reducing damage to adjacent areas	most methods slower than conventional breaking but dependent on hardness and thickness of materials
(c) Precision cutting	(i) diamond sawing	moderate compared with alternatives which require considerable making good	as for rc, plus ability to work in decorated areas and to cut facing materials with structural members	as for rc
	(ii) stitch drilling	expensive		as for rc
	(iii) abrasive discs	cheap	simple tools, minimum preparation, clean cut	slow; only suitable for cutting thin brickwork or chasing concrete or brick
(d) Fixing holes and holes for services	(i) rock drills	cheap	fast. Good quality holes for mechanical fixings max 75mm diameter	through holes may spall at exit
	(ii) diamond drill	moderate	as fast as rock drilling in some materials. No limit to diameter of hole. Neat, no making good	as for rc
	(iii) Thermic lance	moderate	fast. Will penetrate 6m	smoke and fire hazard
4 Material: METALS (ferrous and non-ferrous)				
(a) Crude cutting	(i) explosives	cheap	fast; few restrictions except for licence. Simple, equipment remotely initiated	site evacuation; shock effects
	(ii) Thermit powder	very cheap	no shock or vibration. Simple and remotely initiated	fire hazard
(b) Precision cutting	(i) oxy-acetylene	moderate	skill and equipment readily available	may be slow. Will not easily cut mass iron and non-ferrous metals
	(ii) powder cutting	expensive	will cut most metals in mass	difficult to use on open site conditions in wet
	(iii) Thermic lance	expensive	will cut most metals in mass. Especially effective on heavy, rusty laminations and submerged structures	handling large quantities of lances

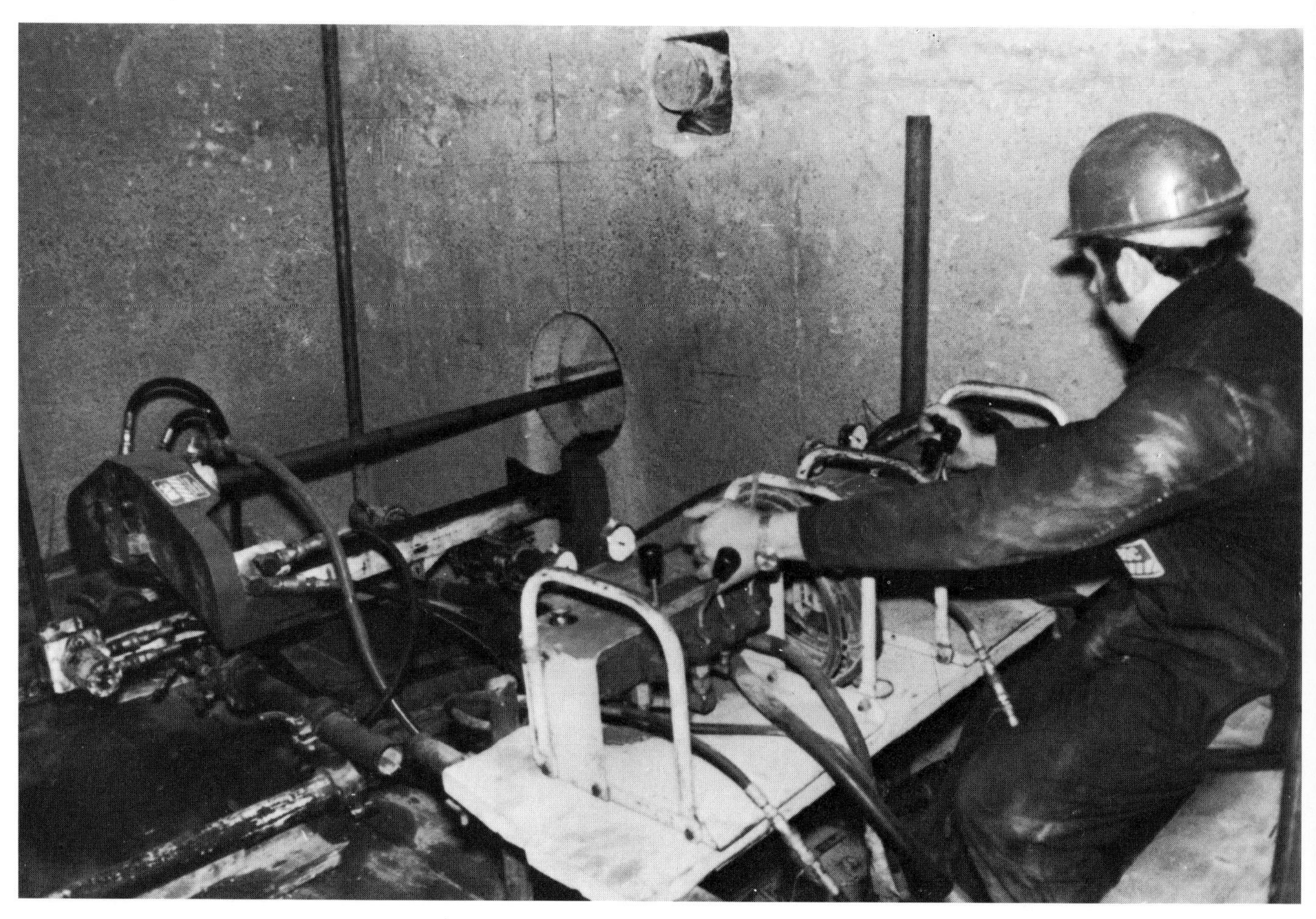

6.13 A specially constructed twin motor, hydraulically operated rig driving a 450mm (18in) diamond core bit, cutting service holes

6.14 Diamond sawn crack control joints in an industrial floor

pre-cast panel and in situ concrete construction. Saw lines can be used to draw the eye away from construction joints, or for purely decorative reasons to give variety to the surface, or may combine decoration and function by forming water sheds, or crack control joints.

The various flame processes also have important decorative applications. A recent programme of experiments with oxy-acetylene flame spalling torches was used to demonstrate the variety of coloured and textured finishes that could be produced by varying the characteristics of the flame, and its "dwell" time on the surface. On the continent of Europe, flame spalling is an accepted method for producing exposed aggregate surfaces on the elevations of buildings. The more powerful flame of the thermic lance or oxy-kerosene torch can be used to burn bas-relief patterns into concrete. High operating costs would make these techniques uneconomical for repetitive designs, but versatile procedures for individual, artistic expression.

The oxy-kerosene torch, in the hands of Edward Monti of Boston, has become a sculptor's tool, with which he fashions superb animal figures and garden ornaments (see pp. 28–29).

7 SPECIFICATION AND ADMINISTRATION

The sequence of steps normally taken in the establishment and management of cutting or breaking, as for any specialized operation, will follow the pattern of:
(i) identifying the problem, and assessing its scope;
(ii) preparing the specification;
(iii) putting the work out for enquiry and quotation;
(iv) placing and administering the contract.
These steps may be simplified or run together, particularly (i) and (ii). The degree of rigorousness of detail may also vary, depending on the complexity or scale of the work. A small amount of straightforward cutting or breaking, which fits easily into the particular physical and contractual situation, will only require the process to be correspondingly brief and simple; on the other hand, a large-scale programme, with one or more difficult cutting operations, possibly to be co-ordinated with a main contractor's working, will require great care in the full specification of the work, the selection of suitable tenderers to submit prices, and informed monitoring of the progress and quality of the work.
Each stage in the procedure will now be reviewed. The following checklist may also be helpful.

Procedure check list

1 Assessment:	Identify problem/need
	Assess scale, quality and scope of work
	Make preliminary selection of potential solutions
2 Specification:	Fully describe objective
	Enumerate the particular requirements
	Identify any constraints
	State contractual conditions
3 Quotation:	Select suitable tenderers
	Obtain prices/rates for the stated scheme or alternative (if available)
	Analyse quotations submitted
4 Contract administration:	Establish correct contract relationships
	Monitor progress and quality of work
	Record and quantify work for payment

7.1 ASSESSMENT

From the guidance given in this book, designers/specifiers should be enabled to appreciate the possibilities of exploiting the techniques which are available. Thus they should aim to be in a position to tackle the work "positively", rather than being driven to consider the whole question only by force of circumstances' and they should aim to have the time to integrate the specialized operations into a broader programme of work, and be able to follow the sequence of actions which have been outlined in a planned way rather than in panic.

The scale of the operations will be apparent from the circumstances which are identified but the necessary quality of work should also be assessed in the light of the objectives. For example, holes to be cut for lifting purposes may not need to be of a highly accurate size or position, whereas holes for installation of services items would need to be more so, and holes for attachment of structural members to an existing structure may need to be very accurate in all respects.
By the use of tables 6.4 and 6.5 the designer/specifier can identify potential solutions to most common situations.

7.2 SPECIFICATION

There are a number of essential elements in specifying this work, in common with many other specialist operations. While the specification notes should be as concise and succinct as possible, they should cover the following items:
To describe the objective fully. The specifier will already have formed an idea of possible methods of operations, but in order to draw on the expertise of the specialist effectively, it is vital that he should be made aware of both the actions to be taken *and* their purpose. Relevant details of the ultimate objective should be clearly stated, eg the type and size of fixing to be inserted in holes, the type of sealant to be applied to saw-cut grooves, the nature of the traffic which is to use a grooved surface, etc.
To enumerate the particular requirements of the operation, in terms of the location of each part of the work, number of items anticipated or areas to be covered, and very importantly the possible *variations* in the work which might be introduced. As the specialist is usually involved in only part of an undertaking with the main contractor or other specialists completing the remainder, the boundaries of these functions and responsibilities must be identified, or clearly left to the main contractor to settle within any necessary design parameters. It is important to specify whether the specialist sub-contractor will have the use of the main contractor's scaffolding, lifting gear, etc, or must provide his own. The specialist will usually price on the basis of having full use of this equipment and he must be informed if it is not available. Expressions such as "normal" services or attendance should be avoided as they are open to differing interpretations.
To identify all known constraints which could influence the work. The constraints might be:
•physical, in terms of access or weight restrictions;
•environmental, in terms of noise, vibration, fumes/dust, or water;
•programme, relative to both hours of working or completion dates.
Limitations are usually imposed on the working hours a specialist contractor may adopt, either to comply with the main con-

tractor's arrangements, because of the pattern of use of an existing building, or to limit the nuisance caused to occupants or local residents. The choice of process, and the type or size of equipment to be used for any particular process, will often depend on these factors. Some research into the physical restrictions may be necessary in order to evaluate floor loading capacities, the sensitivity of buildings or activities in the vicinity of the work, the pattern and timing of these activities which might affect the hours or working, and so on.

Any specification or specification notes, no matter how brief, should set out the contractual conditions which apply. If the work is to be carried out as a sub-contract, the main contract conditions must be identified and any requirements regarding special insurances, etc must be stated. Special instructions as to the basis of pricing and time factors should obviously also be covered.

If the specialist is to be a sub-contractor to a main contractor, the contractor will want to hold his sub-contractor to conditions which interlock with his own. In the UK these will probably be the ICE Conditions, or the Joint Contracts Tribunal (JCT) form (formerly the RIBA Standard Form).

Each will require contract insurances to be effected, the most important of which are:

(a) liability arising from injury to workers;

(b) third party liability for injury or damage to property, arising out of the execution of the works; and

(c) all risks cover for the works, temporary works, and construction plant.

The JCT Standard Form conditions impose an important extension to the contractor's liability to provide cover for damage to any property adjoining that owned by the employer. The nature of many of the cutting/breaking operations could have implications of risk of collapse, or damage to distant areas from flying debris, thus making it necessary to check that the sub-contract works are adequately covered.

It is as well to note that subsidence, collapse, and vibration damage are excluded perils in the wording of the standard third party policy. The risks will have to be examined in the light of data set out in the site engineer's report, in conjunction with a description of any special works, in order to have them included in the contractor's insurance.

7.2.1 EXAMPLES OF SPECIFICATION NOTES

These are included in order to illustrate the way in which the work can be defined, albeit briefly, to cover clearly the objectives, constraints and conditions which apply.

Formation of fixing holes

(i) Fifty fixing holes are to be formed in the existing reinforced concrete frame and infill brick cladding in positions indicated on the drawings. The holes are to receive epoxy resin anchorages for cladding fixings and are to be formed by thermic lance, diamond drilling or other approved non-percussive process.

(ii) The holes are to be left in a suitable condition to meet the requirements of the nominated fixing supplier, Messrs X, who should be approached for any further recommendations regarding the anchorages in this respect. Any necessary secondary drilling or dressing out is to be included in the price for the work, and full details are to be submitted.

(iii) Finished dimensions are to be to an accuracy of +10mm −0 on size, and ±10mm on position.

(iv) Full information on the positioning of ancillary plant is to be submitted before work is begun.

(v) The work will be executed as a nominated sub-contract to the main contractor, under the RIBA standard conditions of contract for nominated sub-contractors. It is anticipated that the work will be complete within one month of date of order.

Enlargement of stairwell by diamond sawing

(i) In order to form a new lift shaft, the existing stairwell is to be enlarged by cutting away the existing landings as shown on the drawings.

(ii) The reinforced concrete slabs are to be cut out to the dimensions shown using a diamond impregnated saw blade in approved equipment. Overcutting of corners will not be accepted, because of the risk of cutting more steel than necessary, but radiusing is allowed to enable corners to be drilled out.

(iii) The building contractor will be responsible for temporary propping of the landings, as approved by the engineer. The specialist sub-contractor is to submit details of the equipment proposed. He will be responsible for ensuring that the equipment does not exceed the maximum safe floor loading, as indicated on the drawings. Details of cooling water and debris drainage, positioning of equipment, temporary support and full information on likely hazards associated with the process are to be submitted before work is begun.

(iv) The work will form a sub-contract to be negotiated with the building contractor, and the specialist is to state what services and attendance he requires for executing his work.

Formation of an opening by stitch drilling

(i) A new doorway is to be formed through the basement wall, in the position indicated on the drawings. The aperture is to be formed by drilling overlapping holes (stitch drilling) and then breaking out the detached panel.

(ii) The access to the work position and the safe floor loading allowance are indicated on the drawings. The contractor should use thermic lancing, diamond drilling or another approved method which does not impart vibration to the structure, and which minimizes noise and dust. Details of the proposed method are to be submitted, with any special requirements for positioning equipment, eliminating fire risk and protection against damage by cooling water or by any other factors associated with the equipment to be used.

(iii) The work will be carried out as a sub-contract with the main contractor, under the RIBA standard form of sub-contract for nominated sub-contractors.

Realignment of basement wall by thermic lancing/water jetting

(i) Part of a reinforced concrete basement wall is to be realigned. The relationship to the street access is indicated on the drawings. The existing basement wall is to be cut to the outline shown by thermic lance (or other proven and approved techniques which impart no vibration to the structure and which minimize noise and dust) and the remainder demolished. After cutting, the reinforcement in the remaining section is to be exposed for a length of one metre and prepared for bonding into the new wall. The exposed reinforcement will be inspected by the engineer for approval.

(ii) Full proposals for handling process water and drainage with details on the positioning of ancillary plant and information on anticipated hazards are to be submitted before work is begun. The concrete is to be removed from the reinforcement by high pressure water jet cutting, or an approved alternative which meets the site restrictions and leaves the reinforcement undamaged.

(iii) The work will be executed as a nominated sub-contract under the RIBA standard conditions of contract for nominated sub-contractors. The work is required to be complete within one month of date of order.

Drilling service holes using diamond drilling

(i) Service holes are required in the reinforced concrete construc-

Date .... Enquiry no. .... **Type of contract**

Customer's name ....

Office address .... Office tel. no. ....

.... Contact ....

.... Site tel. no. ....

Site address .... Contact ....

....

....

Name of contract ....

Date work to be started .... Completed ....

Accommodation for labour .... Rate for accom. ....

Means of travel .... Days of work .... Hours of work ....

Labour force .... Access ....

Limitations on work ....

Facilities to be provided by customer (strike out those unavailable)

PARKING. WEATHER PROTECTION. WATER. AIR. ELECTRICITY. HEATING. SCAFFOLD. STAGING. CARTAGE. TELEPHONE. CANTEEN. FIRST AID. TOILET. WASHING. STORAGE. SAFETY. MARKING OUT. TEMPORARY CONNECTIONS. WATCHING. LIGHTING. EXCAVATIONS. REMOVAL OF DEBRIS. FIRE PRECAUTIONS. WATER/FLUID BARRIERS. PUMPING. DRAWINGS. HOISTING.

Comments ....

Description of work to be done

Customer's drawing no. .... Use overleaf for sketch and additional details

Locations	**Technique**	**Distance from G/L**	**Nature of material**	**Dimensions**	**Time allowance**

Special instructions—Comments ....

Equipment needed

Typical job data sheet of UK cutting contractor

CONTROLLED DEMOLITION INC, Baltimore, Maryland

General information required by CDI **Due date** ....

Inquiry from: .... Phone ....

Name

....

Address

....

Job name: ....

Location: ....

Description: ....

Construction: .... **Steel** .... **Concrete** .... **Wood** ....

Dimensions: ....

Stage of job: ....

Bidding .... **Bids due** ....

Investigating ....

Under contract ....

Time frame: Urgent ....

Funds available for immediate CDI visit to site ....

Look when in area, at no charge ....

Deadline date for CDI input ....

Completion date ....

Nearest airport to job with minimum runway of 2200ft ....

Exposures: Send a plot plan of the area showing scale distances to nearest exposure, and the nature of those exposures ....

.... streets, utilities, or structures

Insurance: What limits of insurance will be required? ....

Others to be named: ....

Photos and prints: Could you send some photographs? .... Polaroid pictures are OK.

Are "as designed" or "as built" prints available? ....

If so, where? ....

Could you send us a copy? ....

Where did you hear about CDI? ....

Use other side for additional information and sketches

Typical job data sheet of explosive demolition contractor in the USA (1)

CONTROLLED DEMOLITION INC, Baltimore, Maryland

Building information required by CDI **Due date** ....

Inquiry from: .... **Phone** ....

Name

....

Address

....

Job name: ....

Location: ....

Description: Type of construction

Reinforced concrete.........., Steel frame.........., Brick-mill (loadbearing brick walls with wood columns and floors)

....

Concrete columns?..........Dimensions or diameter?..........

Steel columns?..........Web and flange dimensions?..........

Are steel columns built up or hot rolled?..........

Distance between columns?..........

Are steel columns concrete, brick, or tile encased?..........

Height of 1 storey, floor to ceiling..........

Does it have a full basement?..........Vaulted sidewalk?..........

What is its depth?..........

What direction do heaviest floor beams run?..........

Floor beams are made of? Concrete.........., Steel.........., Wood..........

Floors are: Concrete slab.........., Thickness.........., Pan construction.........., Wood..........

Use other side for additional information and sketches

Sketch of building to be demolished should include streets and alleys, sidewalks, property lines, utilities poles, fire hydrants, known underground utilities, other buildings to be demolished, and exposure to the demolition, such as adjacent properties; to scale, if possible.

Typical job data sheet of explosive demolition contractor in the USA (2)

Table 7.1 Services and supplies

Operation	Supplier	Characteristics
Major wrecking and clearance	demolition/wrecking contractors	members of the trade association with accepted standards of payment for skills; accept site responsibility and apply necessary safeguards; use traditional methods; generally have heavy transport to remove debris; local operators
Limited breaking concrete and brick, etc (too small for demolition contractors above, but restricted site)	plant hire plus driver	implies tractor-mounted equipment such as hydraulic impact breaker or Nibbler; generally open site conditions; accept no responsibility; local operators
	specialist diamond or lancing contractor with secondary breaking capability	should have equipment and skills in use of wedge bursters, rock drills, and pneumatic breakers; local/national operators
Thermic lancing and powder-cutting	small number of specialist cutting companies	have plant for dispensing large volume of oxygen on site (usually stored in liquid form); safe working requires skill and experience; national operators
Diamond drilling	many small sub-contractors, plus a few larger comprehensively-equipped firms	majority of work can be done with low cost equipment, hence many "one-man bands"; larger jobs may require high cost machines; some contractors specialize in test coring or fixing holes; local/national operators
Diamond sawing	several medium-sized contractors (10–30 employees)	high cost of wall and floor saws restricts the number of suppliers; all offer more than one service, always drilling and secondary breaking, sometimes lancing or powder cutting; local/national operators
Water jet cutting	jet cleaning contractors with high press. pumps; sometimes pump manufacturers	low demand, and service only provided by contractors who happen to have the plant for other uses, eg stone cleaning; accept little or no responsibility; local/national operators
Surface treatment Bush hammering and scabbling	many general purpose contractors	combine several activities, usually including chasing and abrasive cutting; local/national operators
Flame spalling	a few specialist floor treatment contractors; industrial gas suppliers	requires expensive and sensitive equipment, and the handling of large quantities of gas; national operators more typical of continental European situation
Chasing and abrasive cutting	many general purpose contractors	see bush hammering; local/national operators
Explosive breaking or cutting	a few demolition companies a few explosives specialist firms	rarely interested in providing explosive service alone, but may be part of a comprehensive service trade on mystique, personal reputation, and past experience; international operators
Heavy duty burning of steel or cast-iron structures	some demolition welding/flame cutting specialists; lancing specialists	the facilities demanded of the contractor will vary widely according to the size of the job some specialization in underwater work; all thermic lancing firms will tackle metal burning

tion as shown on the drawings. The slabs are of ribbed construction with hollow clay pot infill.

(ii) The holes are to be drilled with hollow cored diamond impregnated bits, used in approved equipment. Drill rigs are to be rigidly mounted and adequately braced to the surrounding structure to maintain the required tolerance on hole diameter of +10mm −0 and on angle drilled to the workface of ±2°.

(iii) Where "through" holes are called for, spalling of the surface surrounding the breakout point is to be avoided.

(iv) The specialist sub-contractor is to submit information on the equipment to be used and details of the proposed methods of water and debris drainage and of support equipment before work is begun.

(v) No building contractor is employed and therefore no attendance is available. The specialist is to price for provision of all equipment and services necessary for his work except mains electricity. The work is to be carried out as a sub-contract to the services installation contractor, Messrs Y, and is to conform to their programme.

Formation of joints in ground slabs

This example is framed as part of a main contract specification.

(i) These ground slabs are to be cast in strips, and power floated/trowelled.

(ii) Longitudinal joints

Longitudinal joints are to be simple butt joints with a joggle (a keyed joint), as shown on the drawings. The tie bars are to be fixed through the side forms, and any small mortar lips on the top edge are to be removed with a carborundum stone before the adjacent strip is cast. Within 24 hours of casting the adjoining bay, 40mm deep saw-cuts are to be made along the joint lines with a suitable machine using diamond tipped blades. The cuts are to be filled with an approved sealant/filler, as instructed, and used strictly in accordance with the manufacturer's instructions. Cuts are not to be formed by hand.

(iii) Transverse joints

Saw-cuts 40mm deep are to be made in the concrete bed within 24 hours of casting. The cuts are to be made by a suitable machine using diamond tipped blades and are to be filled with an approved

sealant/filler, as instructed, used strictly in accordance with the manufacturer's instructions. Cuts are not to be formed by hand.
(iv) The accuracy of alignment of the cuts is to be ±10mm within any 5m length, ±20mm within the length of each joint in any one bay of the building.

7.2.2 ENQUIRY AND QUOTATION SHEETS

As illustrations of the way in which the specialist contractor will follow a similar procedure in defining the work for which he is quoting, two "enquiry/quotation" sheets are reproduced on pp. 71–72; in each case the customer, or the specialist himself if necessary, will complete the questionnaire. The first is the standard sheet used by a British contractor for general cutting work. The others are part of a series used by Controlled Demolition Inc, a company in the USA; each type of construction to be demolished is treated in a corresponding manner.

7.3 QUOTATION

The first step in obtaining prices for the work is to select suitable firms to tender. The importance of dealing with a specialist contractor who has the necessary experience and resources to cope adequately with the work has been emphasized in previous chapters. Table 7.1 gives guidance on the range of situations and contractors which might commonly be encountered.
In seeking prices from tenderers, it is important to establish how the total cost will be built up, if the work is anything but small and simple. The aim must be to have the data necessary to quantify and cost any interruption of the progress of the work, variations of scope or quantity or quality, and any additional services which might be introduced.
It is suggested that the following items should be covered in the quotations for repetitive or variable works:
(i) the establishment charge, for setting up the equipment and labour on site;
(ii) the extra cost of additional visits to site, if the work cannot be covered in a single visit;
(iii) the rate for carrying out the work, based on an accurate quantity, or the most realistic assessment possible if there is any doubt. The units of quantity are commonly:
number: for holes, and well defined "breakthrough" operations
length: for grooving, stitch-drilling, saw-cutting, and all well-defined "severing" tasks
area: for all surface treatments
volume: for demolition or removal of redundant structures
(iv) daywork rates, particularly labour for additional unexpected activities.
For clearly defined cutting or breaking operations, where the exact scope of the work can be established and specified, and where variations will not occur, a single lump-sum price may be adequate.
In modern times, with inflationary price rises such a familiar feature of commercial life, it is essential to state the period for which the prices are to remain fixed, or agree a formula by which increases can be assessed, eg in the UK this could be by reference to the monthly guide to rates in the industry as published by the Department of Trade and Industry.

ASSESSING THE TENDERS/QUOTATIONS

When the competitive tenders/quotations have been received, they should be analysed so as to form a true picture of which one is most likely to prove cheapest. The aspects to be studied are:
(i) Does the submission fully cover the work specified? If not, then the inconsistencies *must* be clarified, and any adjustments made.
(ii) Are there any conditions attached which are at variance with those instructed, and if so, are they unacceptable? Negotiation may succeed in these being deleted.
(iii) Are all the required rates covered? If not, they must be established.
(iv) Using the quoted rates, bring these to a total sum, based on the *anticipated* scope of the work. It is prudent also to calculate a total figure based on both *reduced* and *increased* quantities, in order to assess if a different tenderer would be more economic under those circumstances.
If any alternative schemes are offered, these should be assessed in the same way. Other aspects of the quotations, such as site facilities required, unusual hazards, and timing of the work (both commencement and completion) may also be relevant.
It is helpful if a table is drawn up summarizing the total costs quoted, for the anticipated workload and *also* for a reduced or extended scale, together with a note of times and special conditions. From this an informed selection can then be made.

7.4 CONTRACT ADMINISTRATION

The aim of any contract should be the achievement of the specified objective at the predicted cost to the client, and the administration of the contract, including the monitoring of the work, should be directed to this end.
The elements of this process can be viewed as being:
(i) the establishment of the correct relationships between the parties concerned;
(ii) the monitoring of the progress and quality of the work;
(iii) the definitive assessment of the payments to be made for the work.
Before any work is commenced by the specialist contractor, he must be informed of the "chain of communication" by which he will receive instructions and payments. An early meeting between the main contractor (if there is one) and/or the client, the engineer and/or architect, and the specialist contractor is essential. Such a meeting should resolve and amplify the programme for the work, the methods or processes and plant involved, any particular requirements of the client, and the means and timing of certification of the work.
It is obviously necessary for the designer/specifier to monitor the work in progress, in order to verify that all is being done as it should be done, and to meet any unforeseen circumstances as they arise. This can be achieved by a series of site inspections, and meetings between the parties concerned. It is most important to record in writing any variations to the work, and matters which may affect final payment for the work. Personal recollections of any matters discussed, if not so recorded, are very liable to different interpretation after the event.
The method by which payment will be quantified will have been established at the time of quotation, unless subsequently varied. The work as executed must be recorded in a corresponding manner so that the correct sum can be calculated; the terms of payment will have been established also in the contract documents.

8 CASE STUDIES

CASE STUDIES INTRODUCTION

The following series of case studies has been selected in order to illustrate the potential of the more common processes covered in the book, and to reflect the imaginative ways in which this can be exploited. Of necessity, some processes and techniques are not represented, as so many noteworthy jobs are unfortunately unrecorded. A high proportion of the case studies are concerned with diamond processes, and particularly with diamond sawing. While this is partly due to the success of the diamond industry in publicizing its products and the results of their use, it also reflects the way in which these processes have dominated the market, particularly in the USA.
The processes and tools represented in the case studies are:

1 Slab grooving for safety improvement (see chapter 2.1.5 for a detailed discussion of the technique).
2 Diamond drilling and wall sawing for building preservation (chapter 2.2.3 and 2.2.4).
3 Wall cutting by diamond sawing and drilling (chapter 2.2.3 and 2.2.4).
4 Diamond stitch-drilling through massive concrete (chapter 2.2.3).
5, 6, and **7** Diamond wall sawing for new construction, building improvement, and plant removal (chapter 2.2.4).
8 Diamond sawing for building relocation (chapter 2.2.4).
9 Diamond floor sawing for building alteration (chapter 2.2.4).
10 Precision cutting for refurbishment (chapter 2.2.4).
11 Cutting in an explosion-damaged building (chapter 2.2.4 and 2.3.4).
12 Thermic lancing for building alteration (chapter 2.3.4).
13 Low cost, silent demolition by thermic lance and jacks (chapter 2.3.4 and 3.6).
14 Demolition using Thermit powder (chapter 2.3.5).
15 Bridge demolition using explosives (chapter 3.3.3).
In each case the work is outlined in order to illustrate the background to the work, and the reasons for adopting the procedure and selecting the technique. A brief commentary on the particular merits or disadvantages of the process in relation to the circumstances is added where relevant.

Case study No 1 Slab grooving for safety improvement

Location Premises in Yorkshire and Cambridgeshire, England.
Problem In all three premises, heavily trafficked areas of floor or pavement were dangerously slippery, particularly in wet weather. At a packaging manufacturer's plant, a high quality granolithic floor was being trafficked by heavily loaded, smooth-wheeled fork-lift trucks. Although this was an internal surface, the trucks brought water into the building on their tyres, and various chemical and mechanical treatments had been tried in attempts to eliminate skidding in wet weather.

At a container depot in Leeds, a similar problem was arising on the concrete slabs in the open air; the fork-lift trucks also had to cope with riding up and over bridging plates when loading materials into the containers, requiring good grip.

At the third premises there was a sloping external access ramp between two levels of a cold-store depot in Yorkshire. The concrete slab had been originally cast with a brushed surface texture, but this proved inadequate to enable vehicle wheels to grip properly on the slope during wet weather, particularly when starting or stopping.

In all three cases, it was commercially essential for the work to be carried out quickly, with minimum disruption of normal working, and that it should provide a permanent solution.

Solution In each case, the most economic and effective solution has been to groove the surface mechanically, using Spetin machines made by Errut Products Ltd. These operate by rapidly "flailing" the surface of the concrete with tungsten carbide tipped tines on a revolving drum.

The grooves were cut to a depth of 4–5mm ($\frac{1}{6}$in) at a spacing of 40–50mm ($1\frac{1}{2}$in plus). The average output of the Spetin 12 machines varied from 35–60m² per hr (40–65yd² per hr), depending on the shape and size of the area concerned and the material being cut; on straightforward areas, outputs of over 80m² per hr are claimed.

In all three of these cases, the plant management and labour declare themselves well satisfied with the results.

Comment The situations which here existed are fairly commonplace. The treatment described is an effective and economic answer. There are machines available using diamond blades, which produce similar results.

This study illustrates the advantages which can be obtained from treatment of an unacceptable item, rather than replacing it, to achieve the required standards of performance.

8.1a An Errut Spetin 12 "flail" grooving machine in operation on the loading bay floor

8.1b Fork-lift trucks operating on the concrete surface of the container depot, after treatment by a "flail" grooving machine

8.1c An Errut machine working on the sloped access ramp. The original concrete surface can be seen just ahead of the machine. The whole surface was so treated in 4½ hours

Case study No 2 Diamond drilling and wall sawing for building preservation

Location Long Beach, California, USA.

Problem The Municipal Centre building at Long Beach incorporated a magnificent and historically important mosaic tile mural above the entrance of an auditorium. When the building was scheduled to be demolished, the mural itself was to be saved. It is the largest one-piece mosaic in the world, and was the work of 65 artists, working for almost a year, during the depression period. It measures some 7m wide (24ft) × 12m high (39ft) and was surrounded by a column and arch structure. The back-up wall was of reinforced concrete, with relatively light steel. When it was decided that the mural might be saved, demolition of the building was well advanced, so that conditions of access and facilities were very poor indeed.

Solution Specialist contractor Concrete Coring Co devised a scheme first to protect the mural with a membrane of clear plastic sheets, then provide extra stiffness and strength with a giant steel "C" clamp welded in place around the wall section, and finally severing the mural from the building by saw-cutting. The first cutting operation was to form a slot 450mm high (18in) through the thickness of the wall below the mural, and for the full width. This was achieved by making a horizontal saw cut above and below the zone required and then diamond stitch drilling out the concrete between. This slot enabled the base of the "C" clamp to be installed. Six tiers of scaffolding were then erected. Eight tangential cuts were then made around the top of the mural so as to free it from the arch above. This enabled the vertical steel beams forming the spine of the clamp and the bracing members forming the top to be placed and welded up. The last cutting operation was the vertical sawing down the sides of the mural. Due to press and television coverage, it was decided that the completion of the clamp, the vertical cutting, and the final removal should be completed in two days. Therefore the vertical cuts were made by two teams, one on each edge, using diamond blades which could penetrate the concrete in one pass, for the full thickness which varied between 325mm and 450mm (13–18in). After verifying that the mural was completely freed, and with full television coverage, a giant crane lifted the mural a few inches, brought it 1m (3ft) forward, and then lowered it to the ground. A thorough inspection revealed that no damage had been caused, and the whole panel was then moved to a temporary storage before incorporation in the new buildings on the same site.

Comment This operation, carried out under most difficult conditions, shows how parts of a building can be removed intact for re-use elsewhere. The applications to items of historical or architectural merit are obvious. Careful pre-planning and co-ordination of specialist suppliers and craneage achieved the speed and economy necessary, even acknowledging that there was no other practicable way of saving the mural in this particular case.

8.2a Horizontal slot at foot of mural. The operator is starting vertical cut from base

8.2b Having made the earlier tangential cuts around the top of the mural, diamond saws formed the vertical side cuts

8.2d The mural being lowered to a horizontal position, ready for transport to its new location. The steel strongbacks and the "C" clamps can be seen. The weight of the whole lift was 70 tonnes

8.2c The entire panel was supported by a 150 tonne crane, before the final cuts were made

Case study No 3 Wall cutting by diamond sawing and drilling

Location An airfield in central England.
Problem An existing hangar building, of reinforced concrete construction, was to be converted to office use. This necessitated the provision of windows in one wall which had been almost imperforate.
Solution The architects advising the building owner specified that diamond sawing was to be used, so as to provide a clean and accurate opening into which the windows would fit neatly, and also to minimize the risk of vibration damage to the very large concrete panels which formed the cladding.
In view of the exceptionally large flint aggregate in the concrete, which was difficult to saw economically with air-powered saws. the specialist contractor proposed the use of hydraulically operated automatic-feed diamond drills to stitch drill the horizontal cuts; these surfaces were, in any event, to be masked by weatherings and sills. The diamond saws were then used for the vertical cuts, which were to be left untreated. A total of 50 window openings were formed, amounting to over 250m (800ft) of sawing and stitch drilling.
Comment By intelligent use of alternative equipment, an economical result was achieved, even though more powerful saws could have coped with the hard concrete. The minimal disturbance to the building can clearly be seen in the photographs.
It would certainly have been better had the nature of the concrete been recognized before work commenced; wherever possible with major operations, particularly in geographical areas where it is known that the local concrete is difficult to cut, a small exploratory exercise should be undertaken.

8.3a The diamond-drilling rig in position, forming the overlapping holes at sill level

8.3b Having completed the stitch-drilled cuts at head and sill level, the diamond saw works on the vertical cuts for the window reveals

8.3c Interior view of the completed opening, with the window frame in position

8.3d View of the completed line of window openings; only re-decoration is now required

Case study No 4 Diamond stitch-drilling through massive concrete

Location Shortlands Pumping Station in Kent, England.
Problem The buildings comprising this pumping station were constructed about 100 years ago, and are now listed as being of architectural interest. New, higher capacity pumps were to be installed, requiring an additional 600mm diameter (24in) suction main, and the water authority was faced with the problem of bringing this new pipe into the building without affecting the appearance of the elevations of the building.
To have brought the pipe in above ground level would thus have been unacceptable, so it was decided to tunnel through the foundations. The hole was to be about 1m (39in) in diameter in order to allow the pipe flange to be passed through, and due to the configuration of the walls on the route of the pipe, the length through mass concrete was over 3·5m (11½ft). The hole was centred approximately 1·5m (5ft) below normal ground level.
Solution Bearing in mind the need to limit vibration and impact on the old building, while cutting through the great thickness of concrete involved, the specialist contractor decided to stitch-drill the hole. He adopted two automatic-feed hydraulic rigs, by Victor Products Ltd, driving 100mm (4in) diameter diamond thin-wall bits. By employing this particular equipment, the contractor was able to drill for up to 18 hours per day if necessary, with only one machine-minder on duty at a time. The total length of drilling needed to complete the circumference and isolate the core, was about 122m (400ft) through well-aged concrete containing a high proportion of ungraded flint aggregate. The core then had to be broken up and removed piecemeal, because of the difficult access to the workface, and the great weight of the core as a monolith.

8.4a General view of pumping station, with the sheet-piled excavation on the left

8.4b The multiple-unit drilling rig in position. The pattern of holes already drilled can be seen

Comment The great thickness of concrete involved, the limitation on vibration and the fairly open situation of the work meant that both thermic lancing and diamond drilling could be considered feasible processes. However, in view of the extremely large number of lances which would have been consumed, thermic lancing would have been comparatively uneconomic. By adopting highly mechanized equipment to operate the diamond core-drills, and thus minimizing the labour content, an economic solution was achieved.
The minimal vibration associated with this technique was quite acceptable, and the power plant could be sited outside the building.

8.4c The core concrete having been removed, the lines of the successive diamond-drilled holes can be seen

Case study No 5 Wall sawing for new construction

Location Chicago, USA.
Problem With the technique of tilt-slab construction, where panels for industrial/warehouse buildings are cast horizontally on top of the oversite concrete slab and then tilted up into position, door and window openings are cut to suit the occupier's requirements when the structure is complete. In this case, a new building was to be provided with a series of windows and one doorway, on an echelon plan arrangement.
Solution The specialist contractor, Concrete Drilling & Sawing Co not only formed the openings but also preserved part of one panel for re-use. Diamond blades of 600mm (2ft) diameter were used on a pneumatic powered machine. In each concrete panel, a vertical strip at each edge had to be left in place for support of the upper part of the panel. The window and door frames were then fixed and sealed directly to the cut edges of the wall panels.
Comment Although the tilt-slab construction method has not been used widely in Britain, some of the associated techniques are obviously of much wider application. Even though statutory planning difficulties might preclude the deliberate construction of buildings with minimal perforations, with a view to these being provided later to suit a particular occupier, alterations to buildings must often be achieved. The contrast with the results of traditional methods is apparent.
This technique obviously requires the designers to take account of the possible later perforations, both aesthetically and structurally. Given this preconception, a building can be offered with a range of combinations of windows or doors, loading-bays, etc.

8.5b Exterior view of the two windows

8.5c One of the sections of concrete slab removed from the window positions, further cut down and re-used *as under-sill panel*

8.5a Interior view of two of the window openings, sawn in the lift-slab wall panels. Precast concrete sill units can be seen

8.5d General view of the attractive and useful feature created in the panel wall

Case study No 6 Wall sawing for building improvement

Location Chicago, USA.
Problem An aquarium, featuring elevated tanks with glass sides for public viewing, required repair and modernization. One of the particular problems was that leakage was occurring around the edges of the original glass panels, which had themselves become obscure. Therefore new glass panels were to be installed against the reinforced concrete walls and base of the tanks, together with a new sealant strip.
Solution Having removed the old glass panels, the exposed face of concrete was carefully cleaned. The new sealant strip was to be placed in a groove in the exposed concrete edge. The Concrete Drilling & Sawing Co used a 200mm diameter (8in) diamond grooving blade, on a pneumatic powered machine, which was clamped to the concrete tank structure. A groove was cut to a depth of approximately 12mm ($\frac{1}{2}$in), in the vertical and horizontal edges, and the corners were carefully trimmed-in using a hand-held pneumatic chisel. Great care and accuracy were essential, in order to avoid spalling the exposed edge of the concrete, and so as to provide an accurate bedding for the waterproof seal.
Comment While this was a fairly small project, it well illustrates how specialized cutting techniques can achieve accurate and delicate results. Without these, it is possible that the exposed end of the walls and bottom slabs would have had to be removed, and then re-constructed with the necessary groove; traditional breaking methods would probably have led to local damage to the concrete, which would be unacceptable in a water-retaining structure.

8.6a Close-up of the diamond grooving blade, with the groove formed in the concrete

8.6b End faces of the fish tanks, showing the concrete side walls and base. The projecting steel bolts, used to fix the glass viewing panels, are also being used to locate the machine track

8.6c A groove being carefully formed at the corner by means of a percussion chasing tool

Case study No 7 Wall sawing for plant removal

Location Los Angeles, USA.
Problem An existing multi-storey brewery building contained a number of steel and stainless steel tanks on each floor; these were required for installation in plants elsewhere, while the particular brewery building was itself to be re-used. The tanks weighed between 10 and 20 tonnes each, depending on the material of which they were made, and were approximately 3·5m high × 4·25m wide × 10m long (12 × 14 × 32ft). There were no openings in the walls of the building of a sufficient size to permit the removal of the tanks in one piece, and it would not have been economic to dismantle the tanks for transport in smaller pieces.
Solution The main contractor Westmont Industries teamed with a specialist contractor, National Concrete Sawing Inc, to devise a technique for making fast, precise openings in every panel at each floor level, in one wall of the building. After scaffolding had been set up, NCS started to saw the openings, starting from the top floor and working downwards.
A Drillistics wall saw rig was used, bolted to the external wall, which was 325mm (13in) thick. The operation was carried out in five stages, using diamond blades of 300, 450, 600, 750 and 900mm diameter (12, 18, 24, 30 and 36in), making cuts respectively 50, 150, 250, 300 and 325mm deep (2, 6, 9, 12 and 13in). The blades were powered by a 9 hp air motor rotated at approximately 1000 rpm, and were cooled with water pumped up from ground level. The blades were made by Cushion Cut, and contained De Beers EMB–S natural grit in 20/30 US mesh size.
The reinforced concrete wall contained 16mm ($\frac{5}{8}$in) reinforcing bars at 300mm centres (12in). Because of the presence of the steel, the metal bonding of the diamonds into the blade segments was relatively soft, so that it wore away to continuously expose new diamond cutting edges at an optimum rate. Altogether, approximately 700m (2150ft) of wall was cut, to form 35 openings each approximately 4 × 5m (13 × 16ft). Lift holes were drilled in each panel so that a heavy duty crane could lift each panel out of the wall, and lower it to ground level. The crane could then similarly lift the tanks, which had been jacked across the floors of the building to their lifting position at the edge. The wall panels were broken up by pneumatic breakers before final disposal.
Comment The economy of the procedure and the accuracy of the wall cutting can be judged from the photographs. The structural integrity of the building was in no way impaired, even though virtually all of the non-structural panel concrete was removed. The excellent finished surface of the cut sides of the concrete members enabled the replacement cladding to be installed with no further preparation.

8.7a Wall of the brewery showing seven openings already cut out

8.7b One of the diamond saws in a horizontal cutting position, working on top edge of an opening

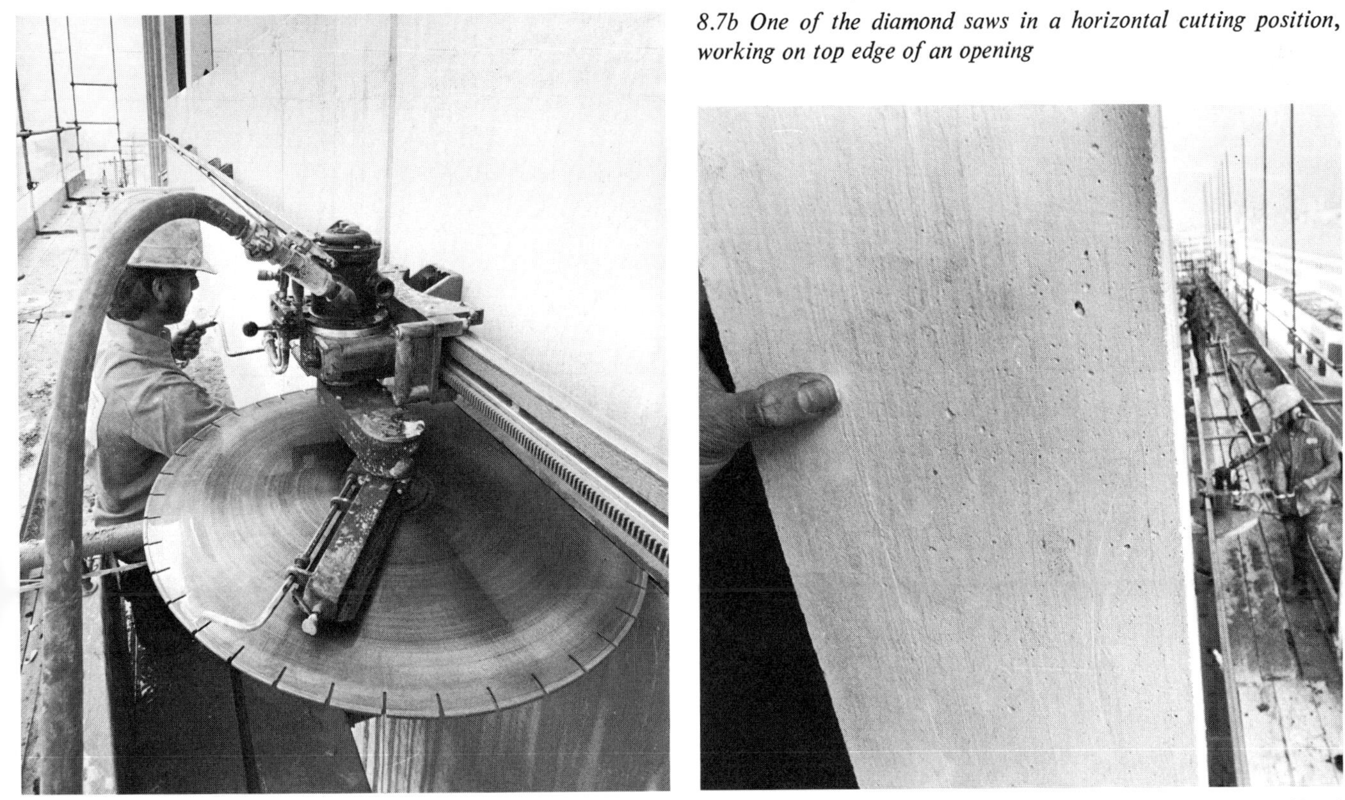

8.7c The diamond blade in operation, with the operator adjusting its cutting depth

8.7d Close-up view of one of the cut faces of the 325mm (13in) concrete walls. The remarkably smooth finish can be appreciated

Case study No 8 Diamond sawing for building relocation

Location Chicago, USA.

1 Problem The old fire station building at Highland Park in Chicago, although a building of considerable local interest, was no longer able to be used for its original purpose on its original site. An alternative social use was found for it, but its new site was some quarter of a mile away.

Solution Specialist contractor Concrete Drilling & Sawing Co carried out the preparatory work to enable the entire building to be freed from its original foundations, and then transported to the new site.

After excavating around the building so as to expose the foundations, a horizontal cut was made through the full width of the foundations along the entire plan length; air-powered diamond saws were used. The building was then jacked up on its supporting grillages, ready for the move.

2 Problem A modern tollway control-room was in similar circumstances to the former example. In order to extend the tollbooth complex, the control room, incorporating a computer room, was to be moved a short distance, but without stopping the control operations *or the computer*.

Solution The same specialist contractor carried out the horizontal concrete sawing so as to sever the building from the foundations, prior to sliding it to its new position. The sawing operations were so carefully executed and screened that the computer suffered no interruption from vibration or dust.

Comment The technique of relocating buildings in their entirety, while not commonly undertaken, is quite well known. These cases illustrate the way in which modern cutting techniques can help to overcome one of the most difficult aspects, that of freeing the building from its old foundations. In case 2, the fact that the computer continued to function throughout is doubly impressive. There are probably many situations where such a concept, if executed as competently as in these cases, could be of wider application. Not only the whole of a building, but perhaps a particular section of, or structure within, a larger building might beneficially be relocated rather than replaced.

8.8a Close-up view of concrete foundation of the Fire House, after it had been completely severed horizontally

8.8b The building ready to be moved to its new location, on a temporary supporting frame

8.8c The building being moved. The horizontal diamond-sawn underside of the foundations can be clearly seen

Case study No 9 Diamond floor sawing for building alteration

Location State of Illinois, USA.

Problem A cold-storage building was to be altered so as to install a new conveyor system between two levels of the building. The floor construction consisted of two thicknesses of reinforced concrete "sandwiching" a layer of insulation material. The total thickness of 500mm (20in) was made up of 150mm concrete above 150mm of insulation above 200mm of concrete (6 + 6 + 8in).

The cold store was not to be interrupted during the work, and therefore the insulation layer was *to remain intact* until the mechanical plant was installed at a later date.

Solution The specialist contractor first erected dust screens around the work area. He diamond drilled a single hole right through the floor, for lifting/supporting tackle; this was the only perforation through the full thickness. He then used air-powered floor saws to cut the concrete to precisely its thickness both above and below the floor, and respectively raised and lowered the severed sections of concrete.

Comment This illustrates well the precision and delicacy which can be achieved with standard equipment, particularly with diamond sawing tools. It is extremely doubtful if any other method could have coped with these most difficult requirements.

8.9a View of the cutting area in the upper storey, with lifting frame in position

8.9b Upper layer of concrete floor being lifted out after cutting

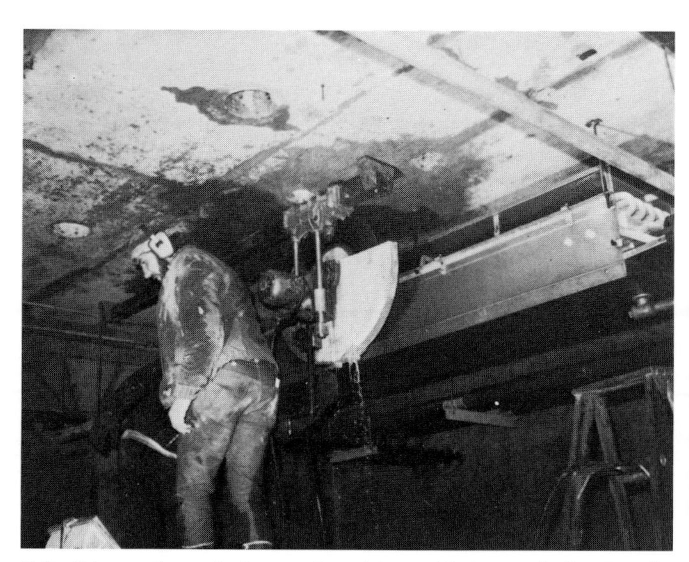

8.9c Diamond saw in inverted position, with its track fixed to the underside of the slab, ready to cut the lower layer

8.9d Lower layer of slab being dropped by cable passed through a diamond-drilled hole in the insulation layer, which can be seen still intact

Case study No 10 Precision cutting for refurbishment

Location Greenwich, England.

Problem Following the local authority's vacation of part of the Greenwich municipal buildings, property developer Town & City Properties embarked on a scheme of refurbishment and extension. This included the insertion of two new floors within the 10m (32ft) high shell of the original council chamber; this was a square building, supported on only four large circular columns, near the corners.

The consulting engineers, Andrews, Kent & Stone, devised an unusual detail to support the new floors on the existing columns, using fabricated steel collars around the reinforced concrete columns. The installation required the cutting of a chase around the columns, approximately 150mm high × 40mm deep (6 × 1$\frac{1}{2}$in) so as to expose but not damage the main vertical reinforcement bars.

Solution Main contractor Sir Robert McAlpine & Sons Ltd first formed two horizontal diamond-sawn cuts, at the top and bottom edge of the chase zone, just to the outer face of the reinforcement, which had been located by covermeter instrument and physical probing. Steel straps were then placed round the concrete, above and below the chase zone, to ensure the integrity of the concrete surface during the operations by minimizing cracking or spalling. Hand-held tools were then used to break out the concrete to form the chase.

After forming the chases in the four columns, the steel collar assemblies were then positioned and bolted, using high strength friction grip bolts. An epoxy-resin grout was introduced into the annular space contained by the collars. After a short time the decking for the new floors was erected and concrete casting could proceed.

Comment This illustrates how an unusual and ingenious structural concept made use of careful cutting techniques with evident success. The necessity for major structural alterations, with minimal damage to existing members (which were to be retained), is of increasing importance. In this case the engineers were able to justify the greatly increased load on the original columns, but this is often not possible; similar techniques may then be required to install strengthening measures, either new members or adding to the existing ones.

8.10a One of the circular columns, with the fabricated steel collar in place. The groove around the column can be seen

8.10b General view of an upper floor with two of the four columns visible. The collars can be seen immediately below the slab

Case study No 11 Cutting in an explosion-damaged building

Location Mersey House, near Liverpool, England.
Problem Mersey House is a multi-storey block of local authority flats, constructed with reinforced concrete frames and floors and infill wall panels of no-fines concrete. A gas explosion in a corner ground floor flat blew out a number of wall panels, cracked the ground floor slab and lifted the first floor slab. Although the structure was still stable in this condition, it was considered that conventional demolition methods carried the risk of extending the damage to other parts of the building. Speed of working and certainty of success were also very important because of the implications of having a large number of people homeless for an indefinite period.
Solution The contractor with responsibility for repairing the property, George Wimpey & Co Ltd, specified that only thermic lancing met the requirements for vibration-free cutting above ground level. Preparatory work on the cracked ground floor slab, mainly consisting of cutting apertures through which new steel supporting columns could be passed to sound foundations, was carried out by diamond sawing.
The work proceeded in a number of stages to allow the erection of a new system of supporting steelwork. Apertures were lanced in the first floor slab to allow the columns to pass through up to the second floor soffit and the strengthening was completed

8.11a The explosion-damaged apartment, showing sections of cladding and structure removed and distortion of edge-beam at first floor level

8.11b Thermic lance at work, removing wall material

8.11c Damaged edge-beam almost separated from the floor, before removal

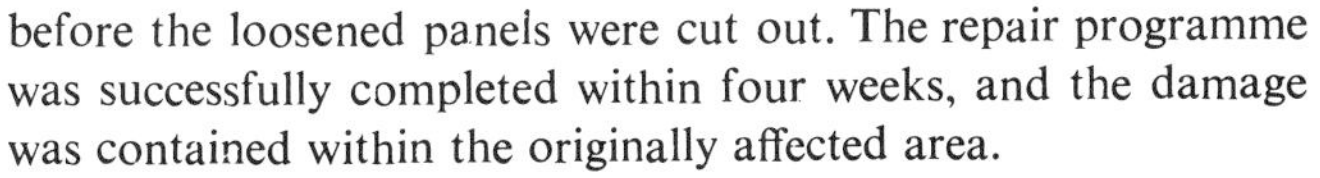

before the loosened panels were cut out. The repair programme was successfully completed within four weeks, and the damage was contained within the originally affected area.

Comment A quick response from the specialist contractor was both required and obtained. The choice of cutting process is interesting, as it illustrates some relative merits of the alternatives. The use of diamond sawing for the formation of holes in the ground slab was predictable; indeed, other than in these very special circumstances, it might well have been the choice throughout. It was the critical necessity for absence of any vibration which led to the use of thermic lancing at higher level. The "open" nature of the premises, at the time of the work, helped in respect of the smoke and fumes which can otherwise be a problem in an occupied building.

8.11d The completed work, after reinstatement of cladding

Case study No 12 Thermic lancing for building alteration

Location London, England.
Problem New transformers were required in the basement of a newspaper plant building. No direct access was available, and it was concluded that a new opening would have to be formed through the 600mm (24in) thick, heavily reinforced floor slab of a passageway, leading from the public street above the basement. The access to this location itself was restricted, and the debris would have to be reduced to no less than six pieces.
Solution The main contractor Trollope & Colls Ltd called in specialist Kaybore Thermic Lancing Co Ltd. It was decided that thermic lancing would be acceptable, as the heavy fumes/smoke could be dissipated to the street at a point where there were no critical environmental factors.
The work area was prepared by clearing all material from the passageway in the vicinity of the work. The basement room beneath, which would be subjected to smoke and sparks, was sealed off and the floor protected with a layer of sand, on asbestos-compound sheeting. Telescopic props were positioned to support each of the severed sections of concrete. The oxygen was supplied from a liquid tanker parked nearby. The cutting operations were carried out by four men in five days, and included the drilling of a central hole through each section of slab to facilitate lifting. A total of 650 lances were consumed.
Comment While the cutting operations themselves were not remarkable, the access and support services and facilities were handled well. Thermic lancing is often rejected as being environmentally unacceptable, but this illustrates that there are circumstances where the objections can be reasonably overcome.
The photographs clearly illustrate the neatness and accuracy of well-prepared cutting work.

8.12a After preliminary clearance of all combustible material from the area and provision of full propping beneath, thermic lancing is commenced

8.12b Two operators working simultaneously

8.12c The manageable sections into which the slab was cut can be clearly seen

8.12d The sections being lifted clear, revealing the propping beneath

Case study No 13 Low-cost silent demolition by thermic lance and jacks

Location The Strand, London, England.
Problem To dismantle two sections of reinforced concrete wall, with its toe slab, which had been constructed to retain temporary access ramps to Charing Cross Station; this was to be achieved without disturbing hotel residents, and the occupants of nearby high-rental offices. Conventional heavy mechanized breaking had quickly run into trouble with local objection on environmental grounds.
Solution The contractors Edmund Nuttall Ltd, working on the new Fleet Line underground station below the Strand, called in specialist contractors.
The concrete was between 350 and 500mm thick, reinforced with 25mm bars at 100mm centres, with some lighter steel of 10mm diameter at 200mm centres. The walls varied in height from 1·5m to approximately 4·6m, and the toe projected horizontally between 2 and 3m from the base.
As speed and low cost were essential, because the need for specialist demolition unfortunately had not been foreseen, the cutting sub-contractor proposed a method using thermic lances to burn out the reinforcing bars only, and to form the pockets for placing heavy duty breaking jacks, which then broke the concrete into manageable pieces. The jacks used were of the type produced by Gullick Dobson Ltd, rated at 80T capacity each, and were used in multiples of three, powered from a common hydraulic pump unit. An electronic covermeter was used to locate the steel bars, and their positions were then marked off for the lance operator to follow. With the strength of the concrete reduced by severing the major reinforcement, the power of the jacks was easily sufficient to accomplish the remainder of the breaking. A total quantity of 20m^3 of concrete was broken into pieces weighing about 2 tonnes each.

Comment The method of working was largely dictated by the necessity of avoiding noise and dust. The choice of thermic lancing which strictly met this requirement was also acceptable in respect of smoke and fumes as the work was in the open air. By minimizing the use of the primary cutting tool, in that the lancing was only used at the positions of the reinforcing bars, the direct cutting costs were reduced to a minimum. As with many such situations, it would have been even more advantageous if the cutting procedure had been pre-planned rather than introduced as an emergency measure.

8.13a The reinforcement is located by means of a "covermeter" instrument

8.13c The jacks inserted into the wall

8.13b Holes are burned by thermic lance, to eliminate the reinforcement, and to accommodate the hydraulic jacks

8.13d The wall fractured at the pre-determined positions, ready for removal

Case study No 14 Demolition using Thermit powder

Location Near Southampton, England.
Problem A wartime defence structure 12m high (40ft), consisting of a reinforced concrete observation post on a lattice steelwork tower, was found to be in a dangerous condition. As the installation was adjacent to Southampton Water, the salt-laden air had corroded the steelwork supports, although the concrete structure above was still intact.
The responsibility for the structure lay with Hamble Parish Council, who had considered alternative methods of demolishing the structure. The site was extremely restricted in area, lying within 3m (10ft) of a newly erected security lodge, alongside the access road to a major hospital. Manual methods of demolition would have required scaffolding and would have been slow and dangerous, but explosives were impractical due to the proximity to the hospital and the risk of damaging the lodge and it was essential to minimize any interruption of access to the hospital.
Solution The specialist contractor who undertook the work, a subsidiary of BOC Ltd, took advantage of the fact that the site was adjacent to the foreshore of Southampton Water and decided that the structure must be dropped in this direction. Thermit charges were used to sever the steel legs, in a predetermined sequence; a section was burned from the two legs nearest to the foreshore, and then the four remaining legs were immediately cut. The charges took about 4–5 hours' work to be placed by three operators, and the actual burning took less than one minute. The road was closed for only about half an hour.
The services of the Fire Brigade were needed, to damp down the small fires which resulted from the white-hot slag falling on to small timber structures in the base of the tower. The debris was broken down into disposable pieces by manual pneumatic tools. The operation consumed about 60kg (125lb) of Thermit powder, at a material cost of approximately £30 ($60) at 1974 prices.
Comment This was a situation where neither conventional means nor explosives could be used for environmental reasons. The only difficulty, in this respect, with Thermit Powder is the volume of smoke produced, but being in the open air this was acceptable.
The solution was extremely economic, and very accurately controlled, and well illustrates the precision which can be achieved with this method.

8.14a Drums of Thermit powder have been attached to the four legs of the structure which are nearest to the foreshore. Two drums have been attached to the leg which is adjacent to the foreshore, so that a length of some 4m of this leg is completely removed initially; these have already been fired

8.14b The remaining drums have now been fired

8.14c The structure collapsing towards the foreshore

8.14d The debris is well contained within the target area. The interruption to use of the adjacent road was only for half an hour

Case study No 15 Bridge demolition using explosives

Location Zanesville, Ohio, USA.

Problem A steelwork pipe-bridge, across a navigable river, was to be replaced. One span, across the main navigation channel, had to be demolished quickly, with minimal obstruction of the channel, and leaving no debris in the river.

Solution Specialist explosives contractor Controlled Demolition Inc of Baltimore first identified the structurally critical members in the girders, which were to be severed; the structure would be explosively cut into a number of sections which would collapse into the river. Each section was securely tied to a buoy, formed of airtight drums; the strength of the steel cable in each case was at least sufficient to support the weight of that particular section of steelwork.

The lay-on explosive charges were attached to the critical members (see p. 38). After firing, the sectionalized steelwork fell into the water, leaving the buoys marking each part. A winch was then attached to each cable in turn, to drag the sections close to the shore, where they could be lifted out by crane. The obstruction to the channel was for a very brief time, and the operation was economically effected.

Comment This was technically straightforward, and careful planning and good engineering sense achieved a remarkably clean and fast operation. There are object lessons to be learnt from this exercise, which can be applied to many other operations, in the benefits to be gained from:

(a) a careful structural assessment to identify the critical members to be cut;

(b) the choice of cutting/breaking method to achieve the instantaneous severence of the members, so that collapse is predictable and not haphazard;

(c) careful assessment of the size/weight of portions of debris, and marking them so that they could be easily recovered.

8.15a View of the bridge, showing the geometry of the members, and the flotation canisters attached to each section of the structure to be separated

8.15b The charges being fired, in this instance simultaneously
8.15c The span falling into the river, with the separate sections now becoming detached

8.15d The broken structure can be recovered, by means of the cables attaching each section to a marker canister/buoy. The recovery crane is visible on the bank

9 MARINE WORK

Marine work and marine structures may call to mind spectacular and highly specialized projects. Britain's North Sea oil exploration and development programmes have also illustrated the severely hostile conditions in which such projects are executed. Nevertheless there are many cutting and breaking jobs in marine work which can be tackled with the same equipment as that reviewed in chapters 2 and 3, or only slightly modified versions of it.

The work may be similar to that on land although marginally more difficult because of the proximity of water, or it may be very different if the work is actually submerged. In the latter case there are many special underwater procedures which are now available, and some of the advanced technology has undoubtedly been stimulated by the North Sea work. It would not be possible to deal fully with these specialized techniques in this book, but it should be useful to examine some of the established cutting and breaking techniques which can be applied to a marine environment.

Work in a marine environment can be classified under three headings in increasing order of severity:

1 above water, where the special factors affecting the work are the organization of services and supplies, and the disposal of debris;

2 at water level, particularly when intermittently flooded (eg within the tidal range), where both the speed of operation and choice of the techniques will be influenced;

3 permanently below water, where the method of operation will be governed by this environment. The difficulty of the work increases with depth, and this review will only be concerned with methods suitable for work down to a maximum of 30m (100ft) depth, beyond which the subject becomes very specialized and beyond the scope of this study.

9.1 WORK OVER WATER

When dealing with work in situations which are themselves dry, but over water, it is this factor alone which distinguishes the conditions from those of a land-based job. While any of the normal techniques can be used, if the necessary support services and facilities can be provided, large masses cannot be dropped to the ground for secondary breaking, and it may not be acceptable to allow debris to remain scattered below water, on the bottom. If pieces of any substantial size have to be dropped into the water, they should have marker buoys attached to lifting points, in order to avoid the costly and time-consuming exercise of exploring the depths with a grappling iron (see Case Study No 12 on p. 96 for an illustration of this). If the pieces are not too large, they may be dropped into nets, or alternatively have flotation buoys attached to them so that they are suspended in the water, which aids recovery.

9.1 Dismantling an rc pile capping beam on a tidal water retaining wall. The broken section of beam is the result of a single small Cardox charge having been released in a hole centred through the beam width. The access staging necessary for working in this environment can also be seen

9.2 Working between tides to thermic lance holes into the granite base of the Longships lighthouse off Cornwall

9.2 WORK INTERMITTENTLY SUBMERGED

Working at any level where the workface is intermittently covered, with a reasonably long period at each state, the contractor may have the option of treating the job as if it were land based or adopting a submersible method. In the first case he will have to tolerate the interrupted pattern of working and, in the second, be prepared to accept the increased costs of working below water. These latter costs will be considerable, and there should be few circumstances where this option would be used.

The working conditions which are to be found close to the water level on exposed sites may be even more arduous than those existing for the diver working well below the surface, due to the effects of wave action. This factor is sometimes overlooked when she difficulties of operations on harbour works and the like are ttudied from the comfort of an office or judged from a site visit on a calm day. The prudent engineer will consult the tables of tidal predictions covering the precise area in which the work is located and will back this up with authoritative local knowledge from coastguards and harbour masters.

For operations which must be carried out underwater, the procedures would be as in 9.3 following.

9.3a Preparing to dive with a thermic lance that has been lit by the attendant on the surface

9.3 UNDERWATER WORK

Nearly all the cutting, drilling, and breaking techniques which have previously been discussed are still applicable, and in some cases they may be found to work even more effectively than in the atmosphere. However, only the properly trained diver should be operating.

The techniques which are most commonly applied to this work are:

- diamond tools (drills and saws);
- flame processes (oxy-acetylene torch and thermic lance);
- water-jet cutting;
- explosives.

9.3.1 DIAMOND TOOLS

Diamond drill bits and saw blades may be used without any loss of cutting efficiency, provided that a suitable power source is made available. Subject to the provision of power, the operation of the tools is very much as it would be on land.

Pneumatic motors are suitable without modification at depths up to 3m (10ft), but allowance must be made for the loss of power due to the back pressure on the exhaust ports; the escaping air also obscures the diver's vision. An all-round improvement can be made by modifying the air motors so that they exhaust into a tube run up to the surface.

Hydraulic motors are usually suitable for underwater use without internal modification since they have a closed circuit system which is effectively sealed. However, the user may have to modify any external bearings or gear trains and repack them with water resistant lubricants if they are to operate continuously at any depth greater than a few feet. The contractor will usually choose to position the power packs for hydraulic equipment above the surface, but submersible self-contained pumps are obtainable, and these are capable of working at depths down to 60m (200ft).

A recent development in underwater power sources is an open loop hydraulic system which uses sea water as the transmitting fluid. This confers the advantage that the fluid can simply be exhausted back into the sea, thereby making a single pipe system that avoids the problem of cooling the return fluid, which has to be effected in most high-power conventional hydraulic systems.

Electrically powered tools are not generally favoured by divers, although a specially designed underwater power system is produced which could be adapted for diamond saws or drills. Electrical equipment can be extremely dangerous if it is used in damp conditions unless the entire system comprising leads, plugs and controls, as well as the motor, have been designed for this duty. Certainly none of the normal range of systems, described as weather-tight, should be used when they are likely to be even momentarily submerged.

9.3b Cutting sheet steel piling below the surface, using a thermic lance

9.3.2 FLAME/HEAT PROCESSES

The underwater burning capability of the conventional oxy-acetylene torch will probably be familiar to the majority of readers; metal cutting by this method can be carried out by

divers in a similar way to that used in dry conditions.
What may be less well known is that the thermic lance shares the same useful attributes. Thermic lancing is effective in shallow water at depths down to about 10m (33ft) using only standard equipment and several diving contractors in the UK are prepared to supply this service. It is usual practice for the attendant on the surface to light the lance, and to pass it down to the diver once it is burning satisfactorily. This, however, obviously creates a protracted work cycle. The lance can be lit underwater using an oxy-acetylene torch which is held by a second diver. Lancing is extremely effective for cutting heavy badly corroded steelwork such as piling, but is less useful on underwater concrete. There is a particular hazard for the diver using the thermic lance, from underwater explosion, which may occur when the lance has penetrated some depth into the material and the escape of the slag and gaseous products of the reaction is becoming restricted. The heat of the reaction in the surrounding water can form pockets of an explosive hydrogen mixture, capable of injuring the diver.
The Clucas lance is an interesting development especially for underwater application. It differs radically from the rigid iron pipe arrangement in that it consists of a plastic-sheathed multi-strand flexible cable with a hollow core; the cable is supplied in 30m (100ft) lengths. Ignition is sustained underwater by striking an arc between the cable end and the workface. Oxygen is passed down the hollow core of the cable, and a 200 amp, 12 volt welding generator is connected across the cable and workface. In this form, since the reaction depends upon having a conducting workpiece, the method is not suitable for concrete. The flexible lance can also be ignited in the usual manner, and will remain self-sustaining below the surface.
It is very effective for cutting badly corroded heavy underwater steel, especially pile joints and laminations where the gaps between the joints usually present a problem to the oxy-acetylene cutter.

9.3.3 WATER-JET CUTTING

The cutting action of a water jet acting on submerged concrete may, under certain circumstances, be more effective than jet cutting in the atmosphere. Experimental work has shown that it is possible to create "cavitation" in the workface which greatly speeds the destructive action. (Cavitation requires the creation of a zone in which evacuated spaces occur within the water flow. They constantly collapse and reform, causing erosion of the surface with which they are in contact.)
Water jets are used below the surface for cleaning and descaling, but they do suffer from the problem of the jet reaction force, and the number of applications for which water-jet cutting would be worthwhile has been too small to support very much development.
Technical advice on the underwater application of water jet equipment is given by such major suppliers of high pressure pumps as F. A. Hughes Ltd and A. Long and Co Ltd in the UK.

9.3.4 EXPLOSIVES

In addition to the expected difficulties of access and handling underwater, the submerged use of explosives is complicated by several other factors, among which are the need to observe special precautions for waterproofing the explosives and firing circuits, and the allowances that have to be made for the overlying weight of water. These are good reasons for employing contractors with substantial previous experience of underwater work, who should be able to obtain the economies of labour and time that the correct use of the technique can often provide.
Although special materials are not always necessary for underwater use at shallow depths, the advantage of using only the marine grades is that if firing is delayed for any reason, everything should remain in working condition until more favourable conditions occur later. The underwater specialist chooses an explosive with a high bulk strength, good water resistance and the ability to resist the reduction in sensitivity caused by hydrostatic pressure.
There are several types and grades of explosive manufactured specifically for submerged applications. The commonly used form is gelatinous explosive such as Subaq 90, or Subgel, as made by the Nobel's division of ICI Ltd; these are highly water resistant, and able to remain in a usable condition at substantial depths for several days in the case of Subaq 90, or weeks for Subgel. An explosive widely used by the offshore oil industry for their particular underwater cutting problems is Xplo Marine Pac, which is a two-liquid component system. The liquids are transported separately, in which state they are completely safe, but when mixed they form an explosive which may be made into shaped charges by pouring into suitable containers.
Detonation can be initiated by electrical impulse or by detonating fuse, the choice depending upon an expert assessment of all the factors of the situation.
Apart from their use on naturally occurring submarine features such as rock outcrops and the deepening of channels, explosives are frequently applied to achieve controlled demolition of all types of structures, and particularly to pile cutting. The design of the structures, and the materials used in construction must first be established by the specialist in order to decide on the best method of applying the explosives. Whenever possible it is preferable to use shot-fired or borehole charges, as opposed to unconfined charges, because the weight of charge needed and the resulting shock wave are both reduced. Consideration of the effects of water-borne shock waves is an important aspect of safety and the contribution that can be made by the expert, in selecting the best approach, can best be seen by comparing the effects of confined and unconfined charges, as stated by ICI:

> A 22·5kg (50lb) charge will produce a perceptible effect in ships up to 185m (200yd) away, whereas borehole charges totalling 810kg (1800lb), and fired in six shots with a short delay between each, produced no noticeable effect at the same distance.

Substantial metal structures, for example caissons or coffer-dams, especially those presenting considerable flat surfaces, can be broken up by utilizing the heavy surge of water following the detonation of the charges to level the plates which the explosions will have loosened.
When controlled localized cutting is specified, this may be achieved by making up lengths of explosive which are placed in immediate contact with the parts to be cut. If the contractor is using blasting gelatine, this will be loaded into canvas or rubberized-fabric tubing which will be held in place on the workface by means of weights, or attached to fastenings. More precision can be obtained by using purpose-made shaped charges but the expertise needed to design and make up the charges is likely to prove expensive. One of the variables that has already been mentioned as affecting the performance of explosives fired underwater is the hydrostatic pressure exerted on the workface. Experienced contractors have the knowledge to assess the amount by which they need to increase the weight of explosive required to achieve the same results at greater depth.
Pile cutting is a frequent marine problem for which explosives are well suited. However, when planning these operations, it is worth giving some thought to tackling them by dry placement of charges, because of the high cost of employing divers. Hollow driven piles, for example, may be cut by lowering the charges down the centre; interlocking piling can be tackled from the

land side by drilling through the back-fill if it is firm consolidated material or concrete.

If the work has to be carried out from the outside, and below the surface, the contractor will probably use submarine blasting gelatine, cartridges of which would be placed end to end in a length of canvas hose, and attached to a line. Two such charges would be made up for each pile to be cut, and one length of each would be made equal to half the circumference of the pile. They would be positioned on diametrically opposing faces of the pile and spaacd slightly apart in the vertical direction. This arrangement when fired would give a good shearing action and result in a clean cut-off. Structures such as bridge piers should be tackled by drilling downwards from the exposed top surface to a level well below the bed of the structure, so that the demolition charges do not leave a residual "stump".

Table 9.1 Examples of cartridge diameter for underwater steel plate cutting, at different depths, using submarine blasting gelatine

Thickness of plate		Diameter of cartridge			Depth of water	
mm	in	mm	in		m	ft
13	½	51	2	less than	10	32
13	½	63	2½		30	100

10 FUTURE DEVELOPMENTS

The techniques that have been described so far in this book are those which have gained sufficient commercial acceptance for both the specialized equipment and the specialist contract skills to be readily obtainable. While it is certain that there is still great scope for further improvement of these techniques, and this is evidenced by the development of the high cycle and hydraulic transmission systems for powering diamond tools, there is also room for radical innovation in both processes and their exploitation. The problem of urban renewal on a massive scale, necessitating the dismantling or major modification of complex structures, is not yet fully upon us but it may highlight the inadequacies of the existing technologies and contractual organization.

10.1 PROCESSES

We are aware that certain theoretical principles have shown promise, under laboratory conditions, of providing the answer to some of the cutting demands which have been forecast, but we can only speculate on which of them, or indeed if any will achieve commercial success. We will review some of the more interesting ones.

10.1.1 THE ELECTROTHERMAL LANCE

Currently under development at the Battelle Research Centre in Geneva is a novel application of a device already widely used in the engineering industry for cutting and welding high melting point metals which is known as the plasma torch. The plasma torch is a heat source created by an electrical discharge of very high power through a gas stream, resulting in temperatures

10.1 The electrothermal lance in its present state of development. The long plasma jet arises principally from the consumption of the electrodes so as to form the nozzle shape. (For reference, the fire-clay bricks in the photograph are 230mm long)

within the discharge zone of 10,000–15,000°C.

In the Battelle device this principle is combined with some of the ideas underlying the thermic lance, to produce a fast and more efficient cutting tool for reinforced concrete. The actual electrothermal lance comprises two co-axial electrodes, the inner one (being the cathode) is the iron tube used to feed oxygen to the burning end, while the outer tube (the anode) is separated from the inner by a space filled with an appropriate flux which stabilizes the arc. A power supply of 40 kW at 110 volts is connected across the electrodes and oxygen is fed through the central core. The combined effects of the high energy arc and the exothermic reaction of the iron tube burning in the oxygen flow will produce a 200mm long plasma jet with a temperature of about 10,000°C. When this is brought into contact with concrete the action is similar to that of the thermic lance in that an iron silicate slag of lower melting point is formed as the burning proceeds. When steel is encountered, this burns in the excess oxygen to aid the combustion process.

It is intended, at a later stage of development, to use the conducting property of the iron silicate slag to enable several lances or subsidiary electrodes to be used on the same workface, with additional electrical power being applied between them. This would have the effect of burning between adjacent lances at the same time as they advance into the material.

The principal advantages claimed for the electrothermal lance over the other flame techniques are lower cost, easier adaptation to vertical or horizontal cutting, and capability of deep cutting to several metres. Although the lances are totally consumed in the process and they will probably be more expensive to produce than the packed thermic lance (which is currently about £4–5 per 3m length) electrothermal lance consumption is low, at about 3m per hour.

10.2 The results of exploratory boring experiments on unreinforced concrete. (For reference, the scale illustrated is 250mm or 10in long)

10.1.2 THE INDUSTRIAL LASER BEAM

A laser is a monochromatic, parallel beam of light which may, or may not, be in the visible spectrum depending upon the method by which it is generated. The laser has many unusual properties, the most important of which, so far as its use as a cutting tool is concerned, is its ability to transmit very high energy with little loss.

When used as a cutting tool, a laser may be reflected from mirrors, and focussed through lenses in the same manner as any light beam until its energy is concentrated upon the non-reflective target. It will then produce intense local heating, depending upon the energy input to the beam, and area of the focus point. When the condition is reached in which the supply of energy to a given area greatly exceeds the rate at which that energy can be dissipated by conduction of heat through the material, then the basis of an efficient cutting action is achieved. At this juncture, work has been principally related to the ability of the laser to apply sufficient local energy to raise the temperature of the workpiece above its melting point. Used in this mode, the material is removed while in its liquid phase, and the heated zone progresses rapidly into the material, along the path of the beam. However, in addition to this mechanism, materials such as concrete, ceramic and natural rock are susceptible to thermal shock; the exploitation of this weakness by the laser beam, to cause the material to break down, may prove to be a more efficient cutting mechanism than simply reducing the material to a molten state.

There are many different types of laser system with widely differing characteristics which suit them to particular applications. The most prominent of the power transmission lasers is the carbon dioxide laser, which is compared with the other systems in **10.4.**

At present, the equipment for producing a laser beam of the

10.3 Industrial laser machine

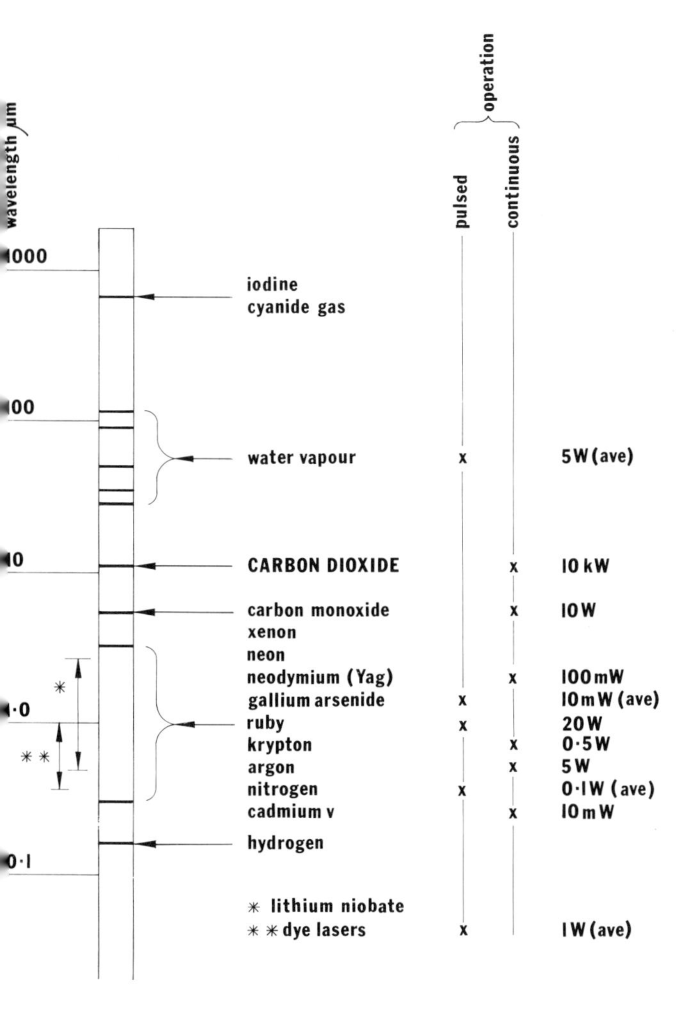

10.4 The range of some of the more important developed lasers. The great power of the carbon dioxide laser in comparison with the alternative active media is evident from the diagram. For this reason it is at present the only significant laser for the cutting industry

order of 10–20 kW is costly (at least £20,000 or $35,000), bulky and unsuited to normal site conditions, so that it can be seen that there is a long way to go before the laser becomes a practical tool and starts to compete with existing procedures.

If these problems are overcome, then it is likely that the laser could be used to create fault lines in the material to be broken, along which breaking would be induced by using a secondary technique. Even then, its advantage over alternative methods seem problematical, but it may be assumed that it will be a quiet, fast, and vibrationless process. The laser does have some inherent dangers, firstly because the radiation may not be visible but still capable of causing severe burns and eye damage, and secondly because of the very high voltages in parts of the equipment.

10.1.3 THE MICROWAVE CONCRETE CUTTER

Electromagnetic radiation of a different kind can be used to produce a heating effect within the body of the material, in contrast to the surface effects resulting fro n the laser beam; such effects are produced by irradiation with microwaves, which are radio waves of high frequency, greater than 10,000 Mhz. Apparatus to produce frequencies of this order can be made to radiate enormous power as may be gauged from the outputs quoted for long-range radar installations which may be up to several megawatts. However, quite modest power levels have been successfully demonstrated to be capable of breaking

10.5 Building Research Establishment's final prototype microwave breaker machine. This was capable of radiating power in the range 2–5 kW

10.6 Crack pattern in oversite concrete slab, developing from point of irradiation

10.7 Concrete slab broken up, after repeated application of the technique described

concrete, and it appears that the optimum return is likely to come from machines of around 10–20 kW capacity.
The mechanism by which breaking is achieved is derived from the internal stresses which arise when the material is locally heated and therefore expands. The components of a multiphase building material respond differently to the radiation. The water content of a cured concrete mixture, which may represent about 8 per cent by volume of the total mass, will absorb the power of the radiation and will turn to steam with a corresponding rise of pressure. The dry materials will also absorb some of the power and expand. When the combined effects produce internal forces in excess of the tensile strength of the material, breakage will occur.
In the series of site experiments undertaken in the UK by the Building Research Establishment, they evolved an interesting technique for using the device. Because the radiation brings about an expanding cone of heated material pointing downwards into the mass, from which cracks radiate into the surrounding cold area, they found that by brushing sand into the cracks that followed the direction of the desired break, it was possible to stop them from reclosing when the power was removed, and the break became self-propagating.
It has already been demonstrated that a practicable, mobile machine can be produced which will give a satisfactory performance; the principle appears to have a number of advantages which may be better exploited in later generations of machines. The method is quiet, imparts no loading on to the workpiece, and is virtually free from dust and fumes. It is however rather slow, but it has been suggested that this may be overcome by setting a number of machines to work simultaneously under one operator, because once they are placed in position and switched on, no further action is required until the heating cycle is completed.

10.1.4 EDDY CURRENT HEATING FOR CUTTING REINFORCEMENT

It is widely appreciated that there are major difficulties in the demolition of prestressed concrete structures. This is particularly true of post-tensioned members in which the tendons are positioned in tubes or ducts, and the high tensile stresses are induced *after* manufacture; after tensioning, the tendons are grouted solid in the ducts (although experience has shown that this is a difficult process, and is often not fully achieved). Any breakage of such a member can lead to sudden and catastrophic failure, which makes demolition both hazardous for the operators and potentially very dangerous in a broader sense in the event of uncontrolled collapse of the building.
One possible solution to the problem is to use eddy current heating to reduce gradually the tension in the tendons, while leaving the concrete undamaged, and enabling the whole structure to be progressively dismantled.
The principle underlying this technique is that when a conductor lies in the path of a fluctuating magnetic field, a current is induced into that conductor which is proportional to the rate of change of the magnetic field. It is quite possible to make this induced current high enough to heat the steel to its plastic state, at which stage the concrete would virtually cease to be reinforced.
At present there are no figures available to indicate the power requirements of such an apparatus, although it is probable that the overall efficiency would be low, due to the difficulties in obtaining a satisfactory magnetic path under practical site conditions.
The components of an eddy current system would typically comprise a diesel driven alternator, a large portable electromagnet or system of magnets, and pole pieces to concentrate the magnetic field.

10.2 EXPLOITATION

10.2.1 IMPROVING EFFICIENCY

In addition to the general improvement of the existing range of equipment, and the introduction of completely new techniques which have just been mentioned, major benefits in terms of faster turnover of work and lower costs will be brought about by bringing modern management methods to bear upon the organization of the specialist contractors, and by a better exploitation of the potential of the techniques by designers and specifiers. In terms of organization, the way ahead has been clearly signposted by a far-sighted Californian contractor who has successfully developed his business philosophy and methods to the point where he is able to market a whole package of equipment plus organization, to other contractors, as a "franchise" type of operation.
It must be recognized firstly that, in general, more time is wasted in loading and unloading the vehicles, and generally putting the operator into a position where he can start cutting, than is taken in actually performing the task; and secondly that the more physically difficult a job is, the less likely it is to be done properly. Edward J. Dempsey set out to correct the situation by evolving a new generation of lightweight high powered drills, saws and ancillary power tools, using compact 400 Hz, three-phase motors. These are used to fit out purpose-built self-sufficient vehicles to form comprehensively equipped and adaptable diamond sawing and drilling units. A great deal of thought has been put into the accessibility and ease of handling of the individual items, both from the point of view of their transit in the vehicle to the site, and their setting up at the workface. The 400 Hz generator can be directly driven through a power take-off on the truck engine, thus avoiding the necessity for a trailer generator. By reducing the weight of the power tools, Dempsey has encouraged the single-handed operator, and by fitting out the vehicles with all the equipment necessary to deal with work from the market which he has first organized to meet, he has dispensed with the time-consuming nonsense of loading and unloading at the depot between each job.
The objective of speeding-up the establishment time for drilling and sawing machines will continue to be a fruitful area for the cost-conscious contractor. Contractors in the USA are much better in this respect than their UK counterparts, and much more use is made of rapid fixings such as cartridge driven studs for locating wall saw tracks, instead of using expanding bolts in holes that have to be laboriously drilled in the workface.
Considerable improvements are equally attainable in the organization and equipping of Thermic lancing units, but it seems likely that these will not be forthcoming unless interest is revived in the wider use of lancing, possibly by reason of other technical improvements. While the lance remains a cumbersome tool to handle, and massive weights need to be transported by reason of the quantities consumed on any sizeable job, it must inevitably decline in use, in favour of more convenient methods.

10.2.2. WIDER APPLICATIONS

Those who are responsible for the design, specification or organization of any work which might involve these techniques will also be able to stimulate the industry. They will do this by a better appreciation of the potential benefits of expanding their use of the processes in conventional work, and also by broadening the horizons of what can be undertaken.
With regard to the broadening horizons of such work, designers and builders should become aware of the facility to treat concrete, brickwork and other hard materials in much the same way as timber has traditionally been treated. As timber members are normally erected or fixed into position and *then* any detailed

cutting, grooving or fixing provisions are made, so concrete and brickwork can be treated. Chapter 6 has already indicated the potential cost advantages of this, quite apart from the reduction of supervisory and management problems which can follow.

Mention has been made in chapter 6 of the concept of erecting a structure with no pre-formed holes for services and fixings; all such holes and pockets are then cut into the structure, assuming of course that the structural design has made allowance for this. Short of this total concept, there are many situations where specialist cutting or breaking is preferable to historic methods. With emphasis on the renewal, alteration or extension of buildings (whether called refurnishment, rehabilitation, or reconstitution in the USA), the scope for the work is enormous, and ever growing. The formation of holes, grooves, or joints using the more sophisticated techniques can be so advantageous that there should be wider acceptance of the role of the specialist contractor.

The potential for removal and/or re-use of parts of buildings, or even whole structures, is illustrated by some of the case studies in chapter 8, particularly Nos 2, 3, 5 and 8. Items of historical or architectural interest, or of economic importance, might be treated in this way. The increasing public sympathy for building and environmental preservation will no doubt be a stimulus in this direction.

The road and pavement industry has likewise found increasing use for mechanised techniques of surface treatment, and cutting of holes for fixings of lighting, balustrades, sign gantries etc. The large machines now available for multiple grooving of concrete and "black-top" surfaces have revolutionized this field of work and the next few years may well see the extension of this to the treatment and improvement of much smaller areas of roads and hard-standings, in factories and industrial estates.

INDEX